COMPUTER NUMERICAL CONTROL

CONCEPTS AND PROGRAMMING
3rd Edition

SOCIETY OF MANUFACTURING ENGINEERS & DELMAR PUBLISHERS INC.

A PARTNERSHIP IN EDUCATIONAL EXCELLENCE

SME and DELMAR have proudly joined forces to form a partnership dedicated to educational excellence. We believe that quality manufacturing education is the key to keeping America competitive in the years ahead.

The Society of Manufacturing Engineers is an international technical society dedicated to advancing scientific knowledge in the field of manufacturing. SME has more than 75,000 members in 70 countries and serves as a forum for engineers and managers to share ideas, information, and accomplishments.

To be successful, today's engineers and technicians must keep pace with the torrent of information that appears each day. To meet this need, SME provides, in addition to the publication of books, many opportunities in continuing education for its members. These opportunities include: monthly meetings through five associations and more than 300 chapters; educational programs including seminars, clinics, and videotapes, as well as conferences and expositions.

Today's manufacturing technology students represent our future. Our goal is to provide these students with the finest manufacturing technology educational products. By pooling our many resources, SME and DELMAR are going to help teachers get the job done.

Together SME and DELMAR will provide outstanding educational materials to prepare students to enter the real world of manufacturing.

Thomas J. Drozda
Director of Publications
Society of Manufacturing Engineers

Joseph P. Reynolds
President and Chief Executive Officer
Delmar Publishers Inc.

COMPUTER NUMERICAL CONTROL

CONCEPTS AND PROGRAMMING
3rd Edition

WARREN S. SEAMES

Delmar Publishers™

I(T)P™ An International Thomson Publishing Company

Albany • Belmont • Bonn • Boston • Cincinnati • Detroit • London • Madrid
Melbourne • Mexico City • New York • Pacific Grove • Paris • San Francisco
Singapore • Tokyo • Toronto • Washington

NOTICE TO THE READER

Dedication

This book is dedicated to my wife, Dolores, for her love, understanding, and patience throughout this project.

Delmar Staff

Senior Administrative Editor: Vern Anthony
Editorial Assistant: Alison Foster
Project Editor: Patricia Konczeski
Production Coordinator: Karen Smith
Art/Design Coordinator: Cheri Plasse

COPYRIGHT © 1995
By Delmar Publishers
A division of International Thomson Publishing Inc.

The ITP logo is a trademark under license

Printed in the United States of America

For more information, contact:

Delmar Publishers Inc.
3 Columbia Circle, Box 15015
Albany, New York 12203-5015

International Thomson Publishing Europe
Berkshire House 168-173
High Holborn
London, WC1V7AA
England

Thomas Nelson Australia
102 Dodds Street
South Melbourne 3205
Victoria, Australia

Nelson Canada
1120 Birchmont Road
Scarborough, Ontario
Canada, M1K 5G4

International Thomson Editores
Campos Ellseos 385, Piso 7
Col Polanco
11560 Mexico D F Mexico

International Thomson Publishing GmbH
Königswinterer Strasse 418
53227 Bonn
Germany

International Thomson Publishing Asia
221 Henderson Road
#05-10 Henderson Building
Singapore 0315

International Thomson Publishing-Japan
Hirakawacho-Kyowa Building, 3F
2-2-1 Hirakawacho
Chiyoda-ku, Tokyo 102
Japan

1 2 3 4 5 6 7 8 9 10   XXX   01 00 99 98 97 96 95 94

Library of Congress Cataloging in Publication Data

Seames, Warren S.
   Computer numerical control: concepts and pro-
gramming/Warren S. Seames.—3rd ed.
      p.   cm.
   Includes index.
   ISBN 0-8273-6498-9.—ISBN 0-8273-6499-7
                                             Instructor's Guide
   1. Machine-tools—Numerical control.   I. Title.
TK3001.E378   1994   9410130
621.9'023—dc20   CIP

CONTENTS

PREFACE

This third edition of *Computer Numerical Control: Concepts and Programming* continues to focus on the basic fundamentals of numerical control (NC) programming. Over the past eight to 10 years, a certain amount of de facto standardization has taken place in the manufacture of NC controls. Although each manufacturer offers a host of different options and capabilities, the fundamental programming codes and syntax required for basic NC programming have become more similar. The core programming chapters have been revised to reflect this trend. All program examples now utilize the coding style used on the more popular current NC controls. New, expanded examples have been added.

The programs represented in this text are not as complex as those found in industry. The examples were carefully selected to demonstrate the basic concepts of CNC programming. The author's goal is to eliminate the learner's confusion that often accompanies the introduction of CNC programming. The learner will achieve an understanding of the programming principles through the programming applications and problems presented in the text. With a firm understanding of the basic principles, the learner can then move confidently ahead to apply the principles to actual industrial situations. Programming is an art, and like art in its visual form, the programming practices used by programmer or NC instructors may vary. However, the basic principles of CNC programming are exhibited in each programmer's effort.

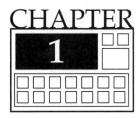

CHAPTER 1

An Introduction to Numerical Control Machinery

OBJECTIVES Upon completion of this chapter, you will be able to:

- Describe the difference between direct and distributive numerical control.
- Describe the difference between a numerical control tape machine and a computer numerical control machine.
- Describe four ways that programs can be entered into a computer numerical controller.
- Explain two tape code formats in use with computer numerical control (CNC) machinery.
- Give the major objectives of numerical control.

Welcome to the world of numerical control. Numerical control (NC) has become popular in shops and factories because it helps solve the problem of making manufacturing systems more flexible. In simple terms, a *numerical control machine is* a machine positioned automatically along a preprogrammed path by means of coded instructions. The key words here are "preprogrammed" and "coded." Someone has to determine what operations the machine is to perform and put that information into a coded form that the NC control unit understands before the machine can do anything. In other words, someone has to program the machine.

Machines may be programmed manually or with the aid of a computer. Manual programming is called *manual part programming;* programming done by a computer is called computer *aided programming (CAP)*. Sometimes a manual program is entered into the machine's controller via its own keypad. This is known as *manual data input (MDI)*. This text will focus on manual part programming.

Advances in microelectronics and microcomputers have allowed the computer to be used as the control unit on modern numerical control machinery. This computer takes the place of the tape reader found on earlier NC machines. In other words, instead of reading and executing the program directly from punched tape, the program is loaded into and executed from the machine's computer. These machines, known as *computer numerical control* (CNC) machines, are the NC machines being manufactured today. *The* primary focus of this text is the MDI programming of *computer numerical control* (CNC) machinery.

THE HISTORY OF NC

In 1947, John Parsons of the Parsons Corporation, began experimenting with the idea of using three-axis curvature data to control machine tool motion for the production of aircraft components. In 1949, Parsons was awarded a U.S. Air Force contract to build what was to become the first numerical control machine. In 1951, the project was assumed by the Massachusetts Institute of Technology. In 1952, numerical control arrived when MIT demonstrated that simultaneous three-axis movements were possible using a laboratory-built controller and a Cincinnati Hydrotel vertical spindle. By 1955, after further refinements, numerical control became available to industry.

Early NC machines ran off punched cards and tape, with tape becoming the more common medium. Due to the time and effort required to change or edit tape, computers were later introduced as aids in programming. Computer involvement came in two forms: computer aided programming languages and direct numerical control (DNC). *Computer aided programming languages* allowed a part programmer to develop an NC program using a set of universal "pidgin English" commands, which the computer then translated into machine codes and punched into the tape. *Direct numerical control* involved using a computer as a partial or complete controller of one or more numerical control machines (see Figure 1-1). Although some companies have been reasonably successful at implementing DNC, the expense of computer capability and software and problems associated with coordinating a DNC system renders such systems economically unfeasible for all but the largest companies.

Recently a new type of DNC system called *distributive numerical control* has been developed (Figure 1-2). It employs a network of computers to coordinate the operation of a number of CNC machines. Ultimately, it may be possible to coordinate an entire factory in this manner. Distributive numerical control solves some of the problems that exist in coordinating a direct numerical control system. There is another type of distributive numerical control that is a spin-off of the system previously explained. In this system, the NC program is transferred in its entirety from a host computer directly to the machine's controller. Alternately, the program can be transferred from a mainframe host computer to a personal computer (PC) on the shop floor where it will be stored until it is needed. The program will then be transferred from the PC to the machine controller.

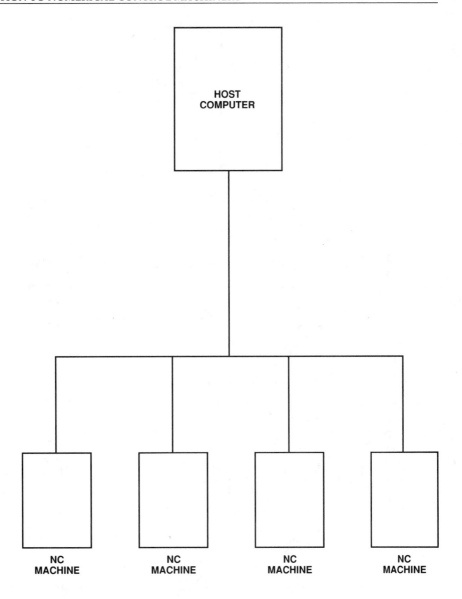

FIGURE 1-1
Direct numerical control

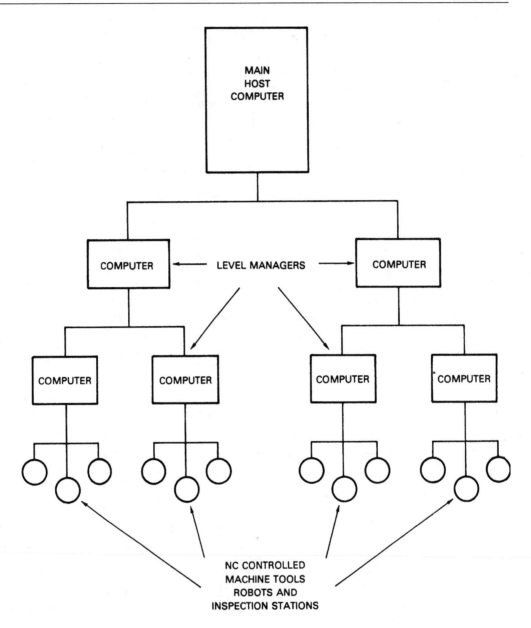

FIGURE 1-2
Distributive numerical control

CNC MACHINES

Figures 1–3 to 1–10 show modern CNC machines. CNC machines have more programmable features than older NC tape machinery and may be used as stand-alone units, or in a network of machines such as a flexible machining center (described in Chapter 16). They are easier to program, and most CNC machines may be programmed by more than one method.

All machines can be programmed via an on-board computer keyboard. In addition to the keyboard, there is a tape reader or electronic connector to allow the transfer of a program written elsewhere to the CNC machine.

A CNC machine is a *soft-wired controller;* that is, once the NC program is loaded into the computer's memory, no hardware is necessary to transfer the numerical control codes to the controller. The controller uses a permanent resident program, called an *executive program,* to process the codes into the electrical pulses that control the machine. The executive program is often referred to as the executive "software," but technically speaking, software is a misnomer. The executive program is more appropriately called *firmware.* In any CNC machine, the executive program resides in ROM memory, and the NC code resides in RAM memory.

ROM stands for *read only memory.* The information in ROM memory is written into the electronic chip and cannot be erased without special equipment. ROM can be accessed by the computer, but cannot be altered. This is why the executive program cannot be erased and is always active when the machine is on.

RAM stands for *random access memory.* RAM can be accessed and altered (written to) by the computer. The NC code is written into RAM by either the keyboard or an outside source. The contents of RAM are lost when the controller is turned off. Many CNC controllers utilize a battery backup system that powers the computer long enough for the program to be transferred (saved) to some storage media in the event of a power loss; other CNC controllers use a special type of RAM called *CMOS memory,* which retains its contents even when the power to the computer is turned off.

INPUT MEDIA

Input media are used to electronically or mechanically store the NC programs until they are needed. A program is simply read from the input medium when it is loaded into the machine. As mentioned previously, there are different methods of inputting the NC code into a controller. Whereas old NC machinery could only read programs from punched tape or a direct numerical control system, CNC machines may possess multiple means of program input.

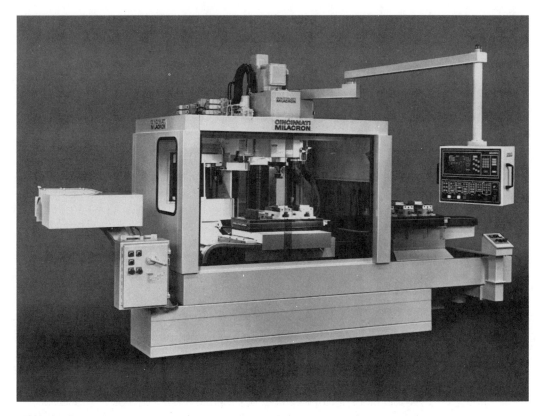

FIGURE 1-3
A vertical spindle machining center, featuring twin pallets *(Photo courtesy of Cincinnati Milacron)*

The oldest medium for program storage is *punched tape* made of paper or mylar plastic (mylar is most commonly used as it is stronger than paper and less likely to tear). The NC program code is entered into the tape by use of a *tape puncher* which punches a series of holes that represent the NC codes. A tape reader employing electrical, optical, or mechanical means senses the holes in the tape and transfers the coded information into the machine computer.

A tape puncher may be attached to a teletype machine or a Flexowriter (a typewriterlike machine). With either of these two pieces of equipment, a code character is punched into the tape as it is being typed onto a sheet of paper. It is more common, however, to see a tape puncher attached to a microcomputer. The NC code is entered into either a CAM (computer-aided manufacturing) or word processor type of program and punched into the tape after all editing of the program is completed.

FIGURE 1-4
A horizontal machining center utilizing twin matrix tool storage magazines. Note the workpiece and tool delivery systems. Safety guards are removed to show clarity. *(Photo courtesy of Cincinnati Milacron)*

Magnetic tape is another popular storage medium. Early experiments with magnetic tape for program storage were not very successful, because the shop environment was not conducive to the delicate tapes of yesteryear. Today's high-quality tapes can survive the rigors of the shop environment with reasonable care in their handling. The most commonly used style of magnetic tape is $1/4$-inch computer grade cassette tape. The cassette case affords good protection and the small size is convenient for storage. Standards for tape format and tape coding have been developed by the Electronics Industries Association (EIA).

BINARY NUMBERS

An understanding of how the controller processes information is helpful in learning to program computer numerical control machinery. Computers and computer-controlled machinery do not deal in Arabic symbols or numbers. All

FIGURE 1-5
A horizontal machining center featuring a Fanuc controller *(Photo courtesy of Bendix Corp.)*

of the internal processing is done by calculating or comparing binary numbers. Binary numbers contain only two digits, zero and one, as illustrated in Figure 1-11 .

Within the CNC controller, each binary digit "one" may represent a positive charge and a "zero" a negative charge, or a "one" may be the presence of charge and a "zero" the absence of charge. The method used depends on the particular controller. In either case, the CNC program code in binary form must be loaded into the computer. Programming formats and languages allow the NC code to be written using alphabetic characters and base-ten decimal numbers. When the NC program is punched or recorded on tape, the information is translated into binary form.

TAPE FORMATS

Various tape coding formats have been developed and tried since the beginning of numerical control. Today, tapes are primarily made using the EIA standard RS-274 format, also called *word address format*. Program information is contained in program lines, called *blocks,* which are punched into the tape in one of two tape code standards. RS-274 is a *variable* block coding

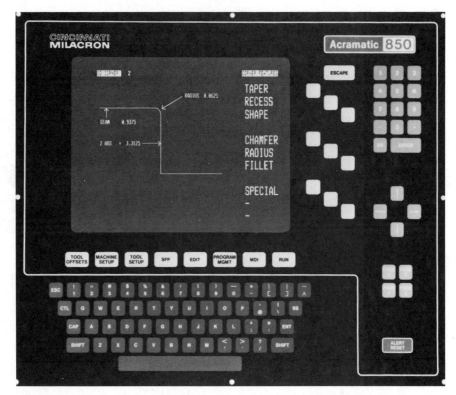

FIGURE 1-6
A CNC controller featuring interactive graphics *(Photo courtesy of Cincinnati Milacron)*

format, meaning that the information contained in a block may be arranged in any order. Discussion of MDI programming using word address format begins in Chapter 6.

RS-244 Binary Coded Decimal

The EIA RS-244 standard, illustrated in Figure 1-12, is one of two tape codes used for NC tapes. It became a standard early in the development of numerical control. Notice that the code utilizes lowercase letters and limited punctuation. Each hole punched into the tape represents the binary digit "one," while a blank space represents the digit "zero." The tape code allows alphabetic characters and base-ten numbers to be translated into the binary code that the controller requires. For this reason, RS-244 is also known as *binary coded decimal* (BCD) .

FIGURE 1-7
A modern vertical spindle machining center *(Photo courtesy of Bridgeport Machines Division of Textron Inc.)*

RS-358

At the time that NC was being implemented in industry, other industries were also using punched tape. Government, telephone, and computer industries all required a tape code that contained both upper and lowercase letters and more punctuation than the limited NC tape code used. What was adequate for numerical control use was not sufficient for other applications. The standard that was adopted for tape coding in these industries was the American Standard Code for Information Interchange (ASCII). To expand the role of computers in NC programming and strive for one standard tape code, EIA RS358 was adopted for use. This code, known as both ISO (International Standards Organization) and ASCII, is a subset of the ASCII code used in other applications. It is illustrated in Figure 1-1 3. Both codes are used to prepare tape for use with CNC machines today. Many CNC controllers can detect which of the two formats is being sent and accept either one.

FIGURE 1-8
A modern horizontal spindle machining center *(Photo courtesy of Cincinnati Milacron)*

OBJECTIVES OF NUMERICAL CONTROL

Numerical control (NC) was developed with these goals in mind:

1. To increase production
2. To reduce labor costs
3. To make production more economical
4. To do jobs that would be impossible or impractical without NC
5. To increase the accuracy of duplicate parts

Before deciding (in light of NC objectives) to utilize an NC or CNC machine for a particular job, the requirements and economics of the job must be weighed against the following advantages and disadvantages of the machinery. Such

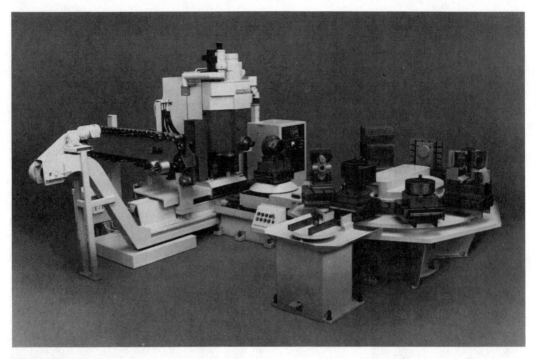

FIGURE 1-9
A horizontal machining center equipped with an eight-pallet automatic workchanger. Safety guards have been removed for clarity. *(Photo courtesy of Cincinnati Milacron)*

an evaluation is necessary to determine if such a machine is practical for the particular job. (Note: NC is a general term used for numerical control. It is also used to describe numerical control machinery that runs directly off of tape. CNC refers specifically to computer numerical control. CNC machines are all NC machines, but not all NC machines are CNC machines.)

Advantages

1. Increased productivity
2. Reduced tool/fixture storage and cost
3. Faster setup time
4. Reduced parts inventory
5. Flexibility that speeds changes in design
6. Better accuracy of parts
7. Reduction in parts handling
8. Better uniformity of parts
9. Better quality control
10. Improvement in manufacturing control

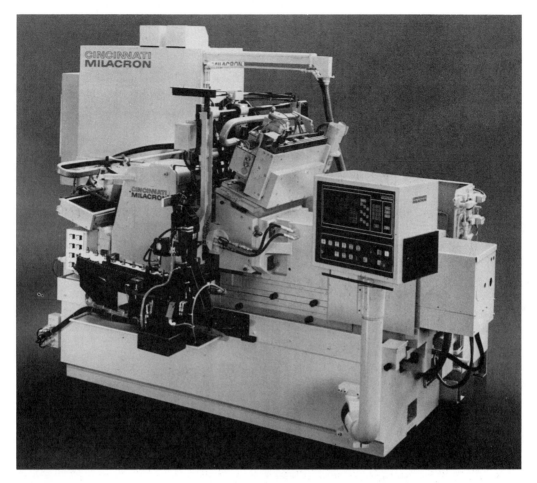

FIGURE 1-10
A CNC centerless grinding machine. This machine features an epoxy granite bed. Safety guards have been removed for clarity. *(Photo courtesy of Cincinnati Milacron)*

Disadvantages

1. Increase in electrical maintenance
2. High initial investment
3. Higher per-hour operating cost than traditional machine tools
4. Retraining of existing personnel

This is not a complete listing of the various advantages and disadvantages of numerical control machines; however, it should give a general idea of the types of jobs for which NC machines are suited.

ARABIC	BINARY	ARABIC	BINARY
0	0	18	1 0 0 1 0
1	1	19	1 0 0 1 1
2	1 0	20	1 0 1 0 0
3	1 1	21	1 0 1 0 1
4	1 0 0	22	1 0 1 1 0
5	1 0 1	23	1 0 1 1 1
6	1 1 0	24	1 1 0 0 0
7	1 1 1	25	1 1 0 0 1
8	1 0 0 0	26	1 1 0 1 0
9	1 0 0 1	27	1 1 0 1 1
10	1 0 1 0	28	1 1 1 0 0
11	1 0 1 1	29	1 1 1 0 1
12	1 1 0 0	30	1 1 1 1 0
13	1 1 0 1	31	1 1 1 1 1
14	1 1 1 0	32	1 0 0 0 0 0
15	1 1 1 1	64	1 0 0 0 0 0 0
16	1 0 0 0 0	128	1 0 0 0 0 0 0 0
17	1 0 0 0 1		

FIGURE 1-11
Binary numbers compared to Arabic numbers

APPLICATIONS IN INDUSTRY

Developed originally for use in aerospace industries, NC is enjoying widespread acceptance in manufacturing. The use of CNC machines continues to increase, becoming visible in most metalworking and manufacturing industries. Aerospace, defense contract, automotive, electronic, appliance, and tooling industries all employ numerical control machinery. Advances in microelectronics have lowered the cost of acquiring CNC equipment. It is not unusual to find CNC machinery in contract tool, die, and moldmaking shops. With the advent of low cost OEM (original equipment manufacturer) and retrofit CNC vertical milling machines, even shops specializing in one-of-a-kind prototype work are using CNCs.

Although numerical control machines traditionally have been machine tools, bending, forming, stamping, and inspection machines have also been produced as numerical control systems. Since this text is written with the student machinist in mind, only CNC machines will be considered.

TRACK NUMBER

8	7	6	5	4	(Feed)	3	2	1	Character
		●			•				0
					•			●	1
					•		●		2
			●		•		●	●	3
					•	●			4
			●		•	●		●	5
			●		•	●	●		6
					•	●	●	●	7
				●	•				8
			●	●	•			●	9
	●	●			•			●	a
	●	●			•		●		b
	●	●	●		•		●	●	c
	●	●			•	●			d
	●	●	●		•	●		●	e
	●	●	●		•	●	●		f
	●	●			•	●	●	●	g
	●	●		●	•				h
	●	●	●	●	•			●	i
	●		●		•			●	j
	●		●		•		●		k
	●				•		●	●	l
	●		●		•	●			m
	●				•	●		●	n
	●				•	●	●		o
	●		●		•	●	●	●	p
	●		●	●	•				q
	●			●	•			●	r
		●	●		•		●		s
		●			•		●	●	t
		●	●		•	●			u
		●			•	●		●	v
		●			•	●	●		w
		●	●		•	●	●	●	x
		●	●	●	•				y
		●		●	•			●	z
	●	●		●	•		●	●	. (Period)
		●	●	●	•		●	●	, (Comma)
		●	●		•		●		/
	●	●	●		•				+
	●				•				-
			●		•				Space
	●	●	●	●	•	●	●	●	Delete
●									CARR. RET. (EOB)
		●		●	•		●		Back Space
		●	●	●	•	●	●		Tab
			●	●	•		●	●	End of Record
					•				Tape Feed Hole
									Blank Tape

FIGURE 1-12
EIA RS-244 tape code

● = Hole in Tape
• = Tape Feed Hole

TRACK NUMBER

8	7	6	5	4		3	2	1	
		●	●		•				0
●		●	●		•			●	1
●		●	●		•		●		2
		●	●		•		●	●	3
●		●	●		•	●			4
		●	●		•	●		●	5
		●	●		•	●	●		6
●		●	●		•	●	●	●	7
●		●	●	●	•				8
		●	●	●	•		●		9
	●				•			●	A
	●				•		●		B
●	●				•		●	●	C
	●				•	●			D
●	●				•	●		●	E
●	●				•	●	●		F
	●				•	●	●	●	G
	●			●	•				H
●	●			●	•			●	I
●	●			●	•		●		J
	●			●	•		●	●	K
●	●			●	•	●			L
	●			●	•	●		●	M
	●			●	•	●	●		N
●	●			●	•	●	●	●	O
	●		●		•				P
●	●		●		•			●	Q
●	●		●		•		●		R
	●		●		•	●	●		S
●	●		●		•	●			T
	●		●		•	●		●	U
	●		●		•	●	●		V
●	●		●		•	●	●	●	W
●	●		●	●	•				X
	●		●	●	•			●	Y
	●		●	●	•		●		Z
●	●	●	●	●	•	●	●	●	Delete
●			●		•				Back Space
			●		•		●		Horiz. Tab
			●		•	●			Line Feed
●			●		•	●		●	Carr. Ret. (EOB)
●		●			•				Space
●		●			•	●		●	%
		●		●	•				((Open Paren.)
●		●		●	•			●	) (Close Paren.)
		●		●	•		●	●	+
		●		●	•	●		●	—
●		●		●	•	●	●	●	/
		●	●	●	•	●			: (Colon)
					•				Tape Feed Hole
									Virgin Tape

FIGURE 1-13
EIA RS-358 tape code

● = Hole in Tape
• = Tape Feed Hole

SUMMARY

The important concepts presented in this chapter are:

- A numerical control machine is a machine that is positioned automatically along a preprogrammed path by way of coded instructions.
- Direct numerical control involves a computer that acts as a partial or full controller to one or more NC machines. Distributive numerical control is a network of computers and numerical control machinery coordinated to perform some task.
- CNC machines use an on-board computer as a controller.
- Offline programming is the programming of a part away from the computer keyboard (hence the term *offline).* This is usually done with a microcomputer.
- There are four ways to input programs into CNC machinery: MDI (manual data input), punched tape, magnetic tape, and DNC (direct numerical control/distributive numerical control).
- Computers work with binary numbers. The CNC program must be loaded into the controller in binary form.
- RS-244 and RS-358 are tape codes used to place information on punched tape. The information is punched into the tape in binary form.
- Before deciding on a numerical control machine for a specific job, the advantages and disadvantages of NC must be weighed in view of the primary objectives of numerical control.

VOCABULARY INTRODUCED IN THIS CHAPTER

ASCII
Binary coded decimal (BCD)
Computer aided programming
Computer numerical control (CNC)
Direct numerical control (DNC)
Distributive numerical control
Input media
Manual data input (MDI)
Manual part programming
Numerical control (NC)
Random access memory (RAM)
Read-only memory (ROM)
Word address format

REVIEW QUESTIONS

1. What is a numerical control machine?
2. What is the difference between an NC tape machine and a CNC machine?
3. What is direct numerical control?
4. What is distributive numerical control?
5. Name four ways to enter a program into a computer numerical control machine.
6. What does CNC mean? What does MDI mean?
7. What are RS-244 and RS-358?
8. What are the five major objectives of numerical control?
9. What are the advantages of NC? What are the disadvantages?

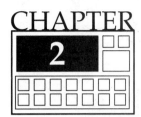

CHAPTER 2

Numerical Control Systems

OBJECTIVES Upon completion of this chapter, you will be able to:

- Describe the two types of control systems in use on NC equipment.
- Name the four types of drive motors used on NC machinery.
- Describe the two types of loop systems used.
- Describe the Cartesian coordinate system.
- Define a machine axis.
- Describe the motion directions on a three-axis milling machine.
- Describe the difference between absolute and incremental positioning.
- Describe the difference between datum and delta dimensioning.

COMPONENTS

A CNC machine consists of two major components: the *machine tool* and the *controller,* or *machine control unit* (MCU), which is an on-board computer. These components may or may not be manufactured by the same company. General Numeric, Fanuc, General Electric, Bendix, Cincinnati Milacron, and G & L Electronics are among those manufacturers of CNC controllers that supply units to makers of machine tools. Figure 2-1 shows a typical controller. Each controller is manufactured with a standard set of built-in codes. Other codes are added by the machine tool builders. For this reason, program codes vary somewhat from machine to machine. Every CNC machine, regardless of manufacturer, is a collection of systems coordinated by the controller.

TYPES OF CONTROL SYSTEMS

There are two types of control systems used on NC machines: *point-to-point systems* and *continuous-path systems. Point-to-point* machines move only in straight lines. They are limited in a practical sense to hole operations

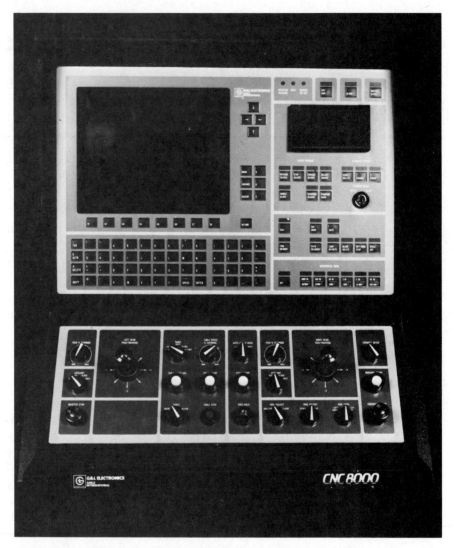

FIGURE 2-1
A modern CNC controller *(Photo courtesy of Giddings & Lewis/Davis Corp.)*

(drilling, reaming, boring, etc.) and straight milling cuts parallel to a machine axis. When making an axis move, all affected drive motors run at the same speed. When one axis motor has moved the instructed amount, it stops while the other motor continues until its axis has reached its programmed location. This makes the cutting of 45-degree angles possible, but not arcs or angles other than 45 degrees. Angles and arc segments must be programmed as a series of straight line cuts (see Figure 2-2).

A *continuous-path* machine (or *contouring* system) has the ability to move its drive motors at varying rates of speed while positioning the machine; the

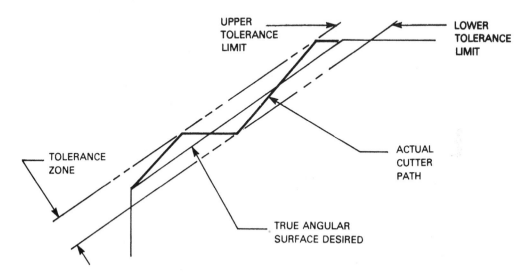

UPPER TOLERANCE LIMIT

LOWER TOLERANCE LIMIT

TOLERANCE ZONE

ACTUAL CUTTER PATH

TRUE ANGULAR SURFACE DESIRED

POINT-TO-POINT ANGLE MADE BY MACHINE
CAPABLE OF 45 DEGREE ANGLES.

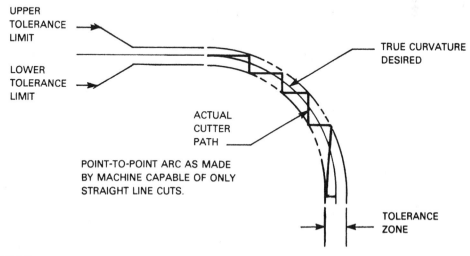

UPPER TOLERANCE LIMIT

LOWER TOLERANCE LIMIT

TRUE CURVATURE DESIRED

ACTUAL CUTTER PATH

POINT-TO-POINT ARC AS MADE
BY MACHINE CAPABLE OF ONLY
STRAIGHT LINE CUTS.

TOLERANCE ZONE

FIGURE 2-2
Point-to-point angles and arcs

cutting of arc segments and any angle may be easily accomplished (see Figure 2-3). At one time, point-to-point machines were common; their electronics were less expensive to produce and they were, therefore, less expensive to acquire. Technological advancements, however, have narrowed the cost difference between point-to-point and continuous-path machines to where most CNC machines now manufactured are of the continuous-path type.

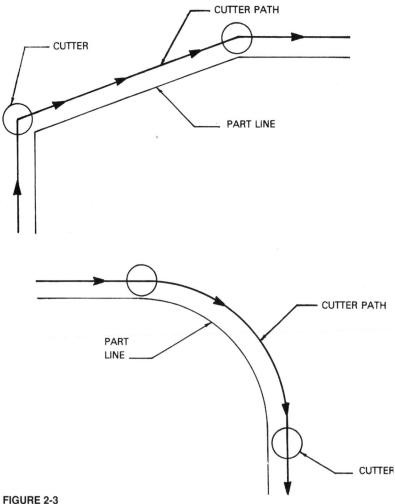

FIGURE 2-3
Continuous-path angles and arcs

SERVOMECHANISMS

It is helpful to understand the drive systems used on NC machinery. The *drive motors* on a particular machine will be one of four types: stepper motors, DC servos, AC servos, or hydraulic servos. Stepper motors move a set amount of rotation (a step) every time the motor receives an electronic pulse. DC and AC servos are widely used variable speed motors found on small and medium continuous-path machines. Unlike a stepper motor, a servo does not move a set distance; when current is applied, the motor starts to turn; when the current is removed, the motor stops turning. The AC servo is a fairly recent development. It can develop more power than a DC servo and is commonly found on CNC machining centers. Hydraulic servos, like AC or DC servos, are variable speed motors. Because they are hydraulic motors, they are capable of producing much more power than an electrical motor. They are used on large NC machinery, usually with an electronic or pneumatic control system attached.

LOOP SYSTEMS

Loop systems are electronic feedback systems that send and receive electronic information from the drive motors. Two types of loop systems are currently in use: open and *closed loop.* The type of system used affects the overall accuracy of the machine. This is valuable to know before selecting a machine to be used for a close tolerance part. Open loop systems use stepper motors; closed loop systems usually use hydraulic, AC, or DC servos.

Figure 2-4 is a block diagram of an open loop system. The machine gets its information from the reader and stores it in the storage device. When the information is needed, it is sent to the drive motor(s). After the motor has completed its move, a signal is sent back to the storage device that the move has been completed, indicating that the next instruction may be received. Notice that there is no process to correct for error induced by the drive system. (There is no such thing as a perfect positioning drive system or motor.)

A closed loop system block diagram is shown in Figure 2-5. As in an open loop system, the machine gets its information from the reading device and stores it in the storage device. When the information is sent to the drive motor(s), the position of the motor is monitored by the system and compared to what was sent. If an error is detected, the necessary correction is sent to the drive system. If the error is large, the machine may simply stop executing the program until the inaccuracy is corrected. This type of system eliminates most errors in position produced by the drive motors.

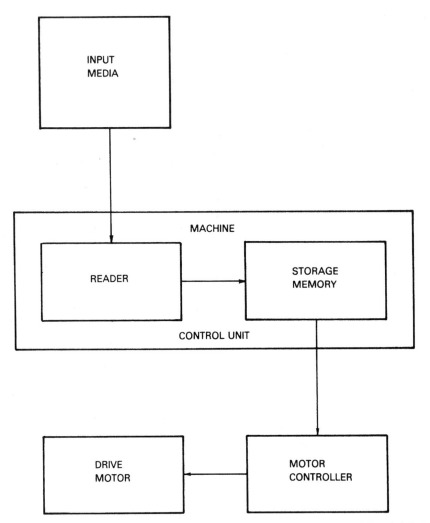

FIGURE 2-4
An open loop system

Recent advances in stepper motor technology have made the manufacture of extremely accurate open loop systems possible. These systems
also eliminate the extra hardware and electronics required for closed loop
systems.

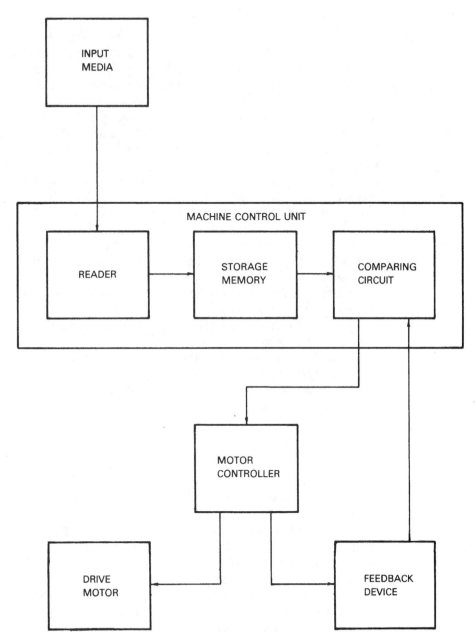

FIGURE 2-5
A closed loop system

THE CARTESIAN COORDINATE SYSTEM

The basis for all machine movement is the Cartesian coordinate system. Figure 2-6 illustrates two- and three-axis coordinate systems. On a machine tool, an *axis is a* direction of movement. The X and Y axes on the coordinate system shown in Figure 2-6(a) can be likened to a two-axis milling machine, where X is the direction of the table travel, and Y is the direction of the cross (or saddle) travel. Figure 2-6(b) illustrates a three-axis coordinate system. Using a vertical mill for example, X would be the table travel, Y the cross (saddle) travel, and Z the spindle travel (up and down). Figure 2-7 illustrates the three axis system on a vertical mill. Machines are also available in four- and five-axis arrangements. A six-axis layout is shown in Figure 2-8. The milling machines programmed in this text will all use this EIA standard axis arrangement.

Cartesian coordinate systems are divided into quarters (quadrants). In Figure 2-9, the quadrants have been labeled I, II, III, and IV, respectively, in a counterclockwise direction. This is the universal way of labeling axis quadrants. Note that the signs of X and Y change when moving from quadrant to quadrant.

Figure 2-10 shows a number of points on a two-axis Cartesian system. Each of the points can be defined by a set of coordinates. The X-axis value is given first; the Y-axis value second. In mathematics this set of points is called an *ordered pair.* In numerical control programming, the points *are* referred to *as coordinates.* In later chapters, Cartesian coordinates will be used in writing numerical control programs.

POSITIVE AND NEGATIVE MOVEMENT

Machine axis direction is defined in terms of *spindle movement.* On some axes, the machine slides actually move; on other axes, the spindle travels. For purposes of standardization, the positive and negative direction of each axis is always defined as if the spindle did the traveling. The arrows in Figure 2-7 show the positive and negative direction of spindle movement along each axis. On a vertical mill, the table would move in the direction opposite to the sign indicated. For example, to make a move in the + X direction (spindle right), the table would move to the left. To make a move in the + Y direction (spindle toward the column), the saddle would move away from the column. The Z axis movement is always positive when the spindle moves toward the machine head and negative when it moves toward the workpiece.

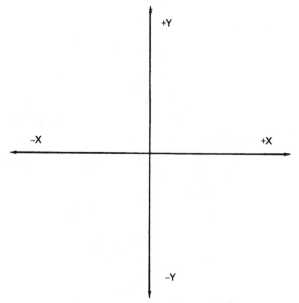

A. TWO-AXIS COORDINATE SYSTEM

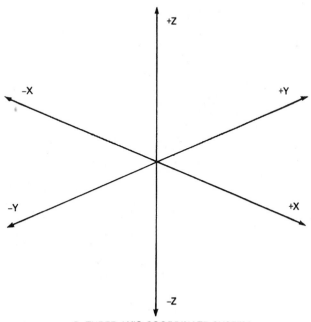

B. THREE-AXIS COORDINATE SYSTEM

FIGURE 2-6
Cartesian coordinate system

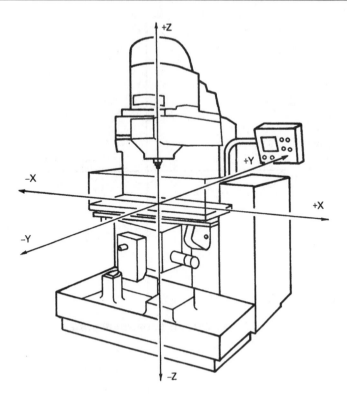

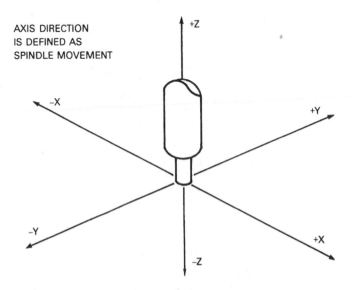

FIGURE 2-7
Three-axis vertical mill

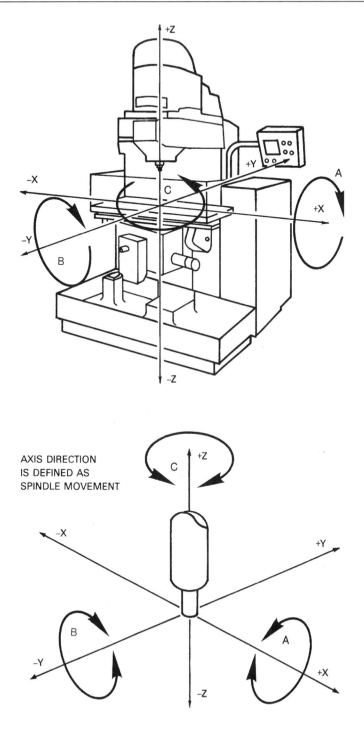

FIGURE 2-8
Six-axis machine layout

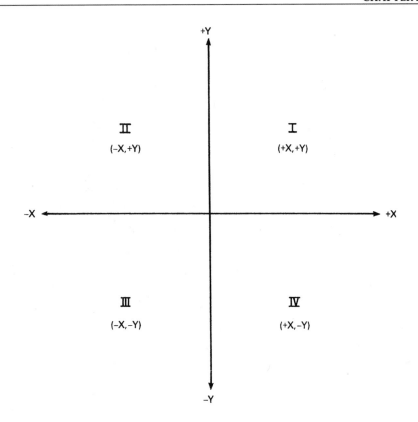

FIGURE 2-9
Cartesian coordinate quadrants

POSITIONING SYSTEMS

There are two ways that machines position themselves with respect to their coordinate systems. These two systems are called *absolute positioning* and *incremental positioning*.

Absolute Positioning

In absolute positioning (Figure 2-11), all machine locations are taken from one fixed zero point. Note that all positions on the part are taken from the X0/Y0 point at the lower left corner of the part. The first hole would have coordinates of X1.000, Y1.000; the second hole coordinates are X2.000, Y1.000; the third hole coordinates are X3.000, Y1.000. Every time the machine moves, the controller references the original zero point at the lower left corner of the part.

Incremental Positioning

In incremental positioning (see Figure 2-12) the X0/Y0 point moves with the machine spindle. Note that each position is specified in relation to the previous one. The first hole coordinates are X1.000, Y1.000; the second hole coordinates are X1.000, Y0.000. The third hole coordinates are again X1.000, Y0.000. After each machine move, the current location is reset to X0/Y0 for the next move. Figures 2-13 and 2-14 illustrate absolute and incremental positioning and their relationship to the Cartesian coordinate system. Notice that with incremental positioning, the coordinate system "moves" with the location. The machine controller does not reference any common zero point.

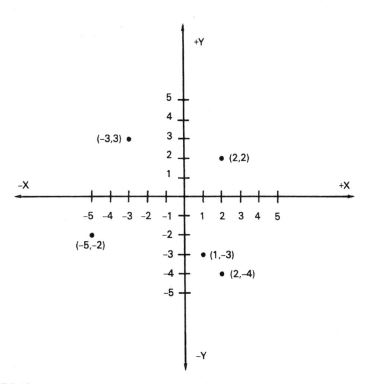

FIGURE 2-10
Cartesian coordinates

SETTING THE MACHINE ORIGIN

Most CNC machinery has a default coordinate system the machine assumes upon power-up, known as the *machine coordinate system.* The origin of this system is called the *machine origin* or *home zero location.* Home zero is usually, but not always, located at the tool change position of a machining center. A part is programmed independent of the machine coordinate system. The programmer will pick a location on the part or fixture. This location becomes the origin of the coordinate system for that part. The programmer's coordinate system is called the *local* or *part coordinate system.* The machine coordinate system and the part coordinate system will almost never coincide. Prior to running the part program, the coordinate system must be transferred from the machine system to the part system. This is known as *setting a zero point.*

There are three ways a zero point can be set on CNC machines: manually by the operator, by a programmed absolute zero shift, or by using work coordinates.

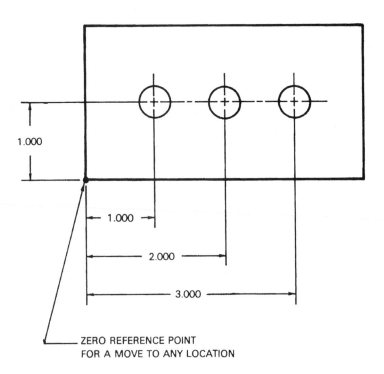

FIGURE 2-11
Absolute positioning

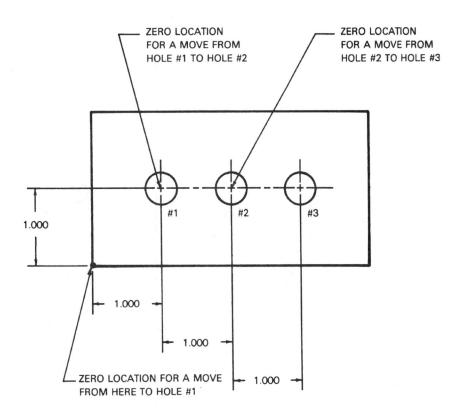

FIGURE 2-12
Incremental positioning

Manual Setting

When *manual zero setting* is used, the setup person positions the spindle over the desired part zero and zeros out the coordinate system on the MCU console. The actual keystroke sequence for accomplishing this varies from controller to controller.

Absolute Zero Shift

An *absolute zero shift is* a transferring of the coordinate system which is done inside the NC program. The programmer first commands the spindle to the home zero location. Next, a command is given that tells the MCU how far from the home zero location the coordinate system origin is to be located. An absolute zero shift is given as follows:

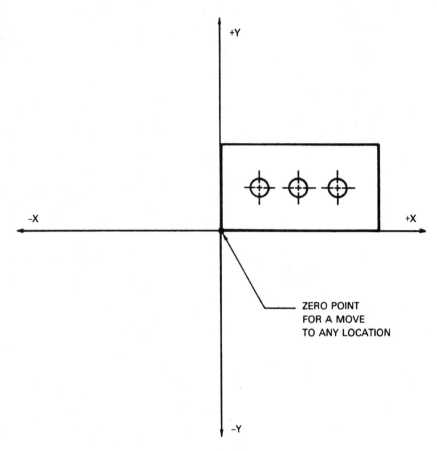

FIGURE 2-13
Relationship of the Cartesian coordinate system to the part when using absolute positioning

```
(Send the spindle to home zero)
N010 G28 X0 Y0 Z0
(Set the current spindle position)
(To X5.000 Y6.000 Z7.000)
N020 G92 X5.000 Y6.000 Z7.000
```

In line N010 the spindle moves to home zero. Following line N020, even though the spindle did not physically move from home zero, the location of the spindle became X5.0 Y6.0 Z7.0 as far as the MCU is concerned. The machine will now reference the part coordinate system. G92 is a fairly standard command for an absolute zero shift. The term G92 line is often used to describe an absolute zero shift.

If more than one fixture is to be used on a machine, the programmer will want to use more than one part coordinate system. By sending the spindle back to home zero using a G28 X0 Y0 Z0 command, another G92 line can be used in the program to set the second part coordinate system.

Work Coordinates

A *work coordinate* is a modification of the absolute zero shift. Work coordinates are registers in which the distance from home zero to the part zero can be stored. The part coordinate system does not take effect until the work coordinate is commanded in the NC program.

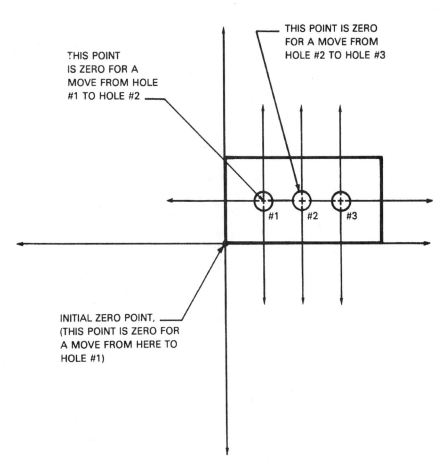

FIGURE 2-14
Relationship of the Cartesian coordinate system to the part when using incremental positioning

When using G92 zero shifts, the coordinate system was changed to the part coordinate system when the G92 line was issued. When using work coordinates, a register can be set at one place in the program and called at another. If more than one fixture is used on a machine, a second part zero can be entered in a second work coordinate, and called up when needed. The work coordinate registers can be set either manually by the operator, or in the program by the NC programmer, without having to send the spindle to the home zero location. This saves program cycle time by eliminating the moves to home zero in the program.

Work coordinates are set and called up in a program by commands called *G-codes.* G54, G55, and G56 would be examples of G-codes to call up different work coordinate registers. The following is an example of using work coordinates:

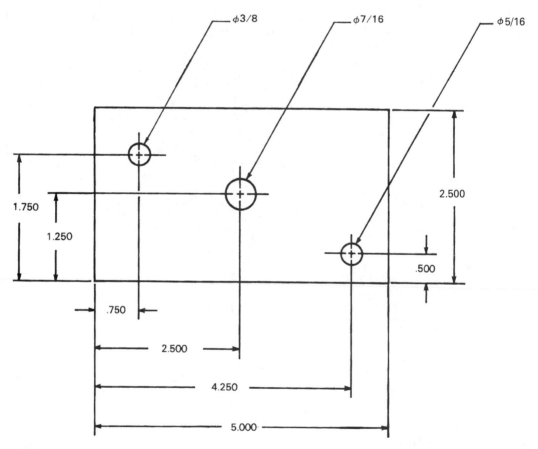

FIGURE 2-15
A datum dimensioned drawing

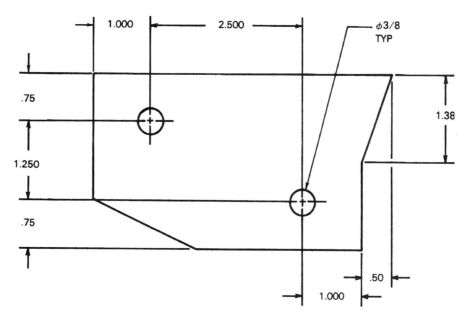

FIGURE 2-16
A delta dimensioned drawing

(Set work coordinate P1—which is G54)
(and work coordinate P2—which is G55)
N010 G10 L2 P1 X5.000 Y6.000 Z7.000
N020 G10 L2 P2 X10.000 Y3.000 Z15.000

(Call work coordinate G54 and move)
(To X1.000 Y1.000 Z.500)
N100 G54 X1.000 Y1.000 Z.500

(Call work coordinate G55 and move)
(To X2.000 Y2.000 Z3.000)
N110 G55 X2.000 Y2.000 Z3.000

In line N010, the G54 work coordinate is set to X5.0, Y6.0, Z7.0 from the home zero location. In line N020, the G55 work coordinate is set to X10.0 Y3.0 Z15.0 from home zero. In line N100, the G54 work coordinate is called, activating the part coordinate system. The spindle is moved to X1.0 Y1.0 Z.5 as referenced from the activated part coordinate system. In line N110, the G55 work coordinate is called, activating the second part coordinate system. The spindle is moved to X2.0 Y2.0 Z3.0 as referenced from the second part coordinate system.

Work coordinates remain active once called until cancelled by another work coordinate. They may be called on a line by themselves, or in a line with motion commands as in the example.

DIMENSIONING

In conjunction with NC (or N/C) machinery, there are two types of dimensioning practices used on part blueprints: *datum* and *delta.* These two dimensioning methods are related to absolute and incremental positioning. (Note: although this text uses the NC abbreviation, N/C is equally accepted and is beginning to become the more prominent form.)

Datum Dimensioning

In datum dimensioning, all dimensions on a drawing are placed in reference to one *fixed* zero point. Datum dimensioning is ideally suited to absolute positioning equipment. Figure 2-15 shows a datum dimensioned drawing; notice how all dimensions are taken from the corner of the part.

Delta Dimensioning

Dimensions placed on a delta dimensioned drawing are "chain-linked." Each location is dimensioned from the *previous* one, as shown in Figure 2-16. Delta drawings are suited for programming incremental positioning machines.

In many cases, the drafting practice does not suit the available machines. It is often necessary to calculate program coordinates from print dimensions because a delta dimensioned drawing is being used to program an absolute positioning machine, and vice versa. It is not uncommon to find the two methods mixed on one drawing.

SUMMARY

The important concepts presented in this chapter are:

- There are two types of NC control systems: point-to-point and continuous path.
- There are four types of drive motors used on NC equipment: stepper motors, AC servos, DC servos, and hydraulic servos.
- Loop systems are electronic feedback systems used to help control machine positioning. There are two types of loop systems: open and closed. Closed loop systems can correct errors induced by the drive system; open loop systems cannot.

- The basis of machine movement is the Cartesian coordinate system. Any point on the Cartesian coordinate system may be defined by X/Y or X/Y/Z coordinates.
- An absolute positioning system locates machine coordinates relative to a fixed datum reference point.
- In an incremental positioning system, each coordinate location is referenced to the previous one.
- The machine coordinate system can be transferred to the part coordinate system manually, by an absolute zero shift, or by use of work coordinates.
- The positive or negative direction of an axis movement is always thought of as spindle movement.
- Machine movements occur along axes which correspond to the direction of travel of the various machine slides. On a vertical mill, the Z axis of a machine is always the spindle axis. The X and Y axes of a machine are perpendicular to the Z axis, with X being the axis of longer travel.
- There are two dimensioning systems used on part drawings intended for numerical control: datum and delta. Datum dimensioning references each dimension to a fixed set of reference points; delta dimensioning references each dimension to the previous one.

VOCABULARY INTRODUCED IN THIS CHAPTER

Absolute positioning
Absolute zero shift
Cartesian coordinate system
Closed loop system
Continuous path
Datum dimensioning
Delta dimensioning
Incremental positioning
Machine control unit (MCU)
Point-to-point
Open loop system
Work coordinates

REVIEW QUESTIONS

1. What is the difference between point-to-point and continuous-path systems?
2. What are the four types of drive motors used on numerical control machines?
3. What is a loop system?
4. What is the difference between an open and closed loop system? Why is the choice of loop system important?
5. What is a machine axis?
6. What machine feature determines the positive or negative direction of an axis?
7. What are the two types of positioning systems? What are the differences between them?
8. What three methods are used to origin a part on a CNC machine?
9. What two types of dimensioning systems are used on NC part prints?
10. Give the coordinates of the points shown in Figure 2-17.

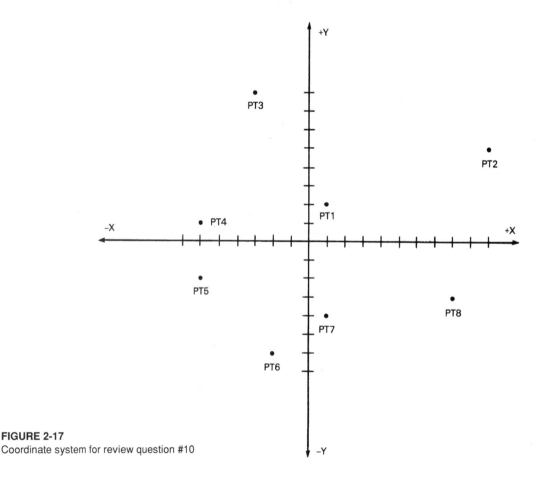

FIGURE 2-17
Coordinate system for review question #10

CHAPTER 3

Process Planning and Tool Selection

OBJECTIVES Upon completion of this chapter, you will be able to:

- List the steps involved in process planning.
- List the factors that influence the selection of an NC machine, workholding devices, and tooling.
- Describe the types of tools available for hole operations.
- Describe the types of tools available for milling operations.
- Determine the proper grade of carbide insert for a given material.
- Describe some common NC turning tool types.
- Determine the proper spindle RPM to obtain a given cutting speed.
- Explain the importance of proper feedrates.

PROCESS PLANNING

Process planning is the term used to describe the development of an NC part program. A number of decisions must be made by the NC programmer to successfully program a part.

- Which NC machine should be used?
- How will the part be held in the machine?
- What machining operations and strategy will be used?
- What cutting tools will be used?

This process is known as *methodizing*—developing the entire method of producing the part.

Machine Selection

A programmer must first decide which machine will be used. This decision is based on a number of factors.

- What is the programmer's experience?
- What machines are available?

- How many parts are in the order? Are there enough to justify the setup time and higher per hour run cost on a more complex machine?
- Is the particular part best suited for a lathe or a milling machine application?
- Is a vertical or horizontal spindle preferred? Vertical spindles are advantageous for hole drilling and boring operations. Horizontal spindles are best for heavy milling operations. The horizontal orientation of the spindle causes the chips to fall away from the tool, whereas vertical spindles tend to keep the chips packed around the tool.

Fixturing

The next decision to be made is how will the workpiece be held? Again this decision is based on a number of factors, many of them economic.

- Will standard holding devices (clamps, mill vises, chucks, etc.) suffice, or will special fixturing need to be developed?
- What quantity of parts will be run? A large number of parts means special fixturing to shorten the machining cycle may be feasible, even if conventional workholding methods would otherwise be used.
- How elaborate does the fixturing need to be? If many part runs are foreseen, a more durable fixture must be designed. If only one or two part runs are projected, a simpler fixture can be used.
- What will make the best quality part?

Machining Strategy

The machining strategy must be developed before the NC program can be written. Machining sequences used in a part program are determined by the following decisions.

- What is the programmer's experience?
- What is the shape of the part and the blueprint tolerance?
- What tooling is available?
- How many parts are in the order?

Tool Selection

Tool selection is the final important step in process planning. The selection is based on the following decisions.

- What tools are available?
- What machining strategy is to be used?
- How many parts are in the order? If a large number of parts are in the order, special timesaving tools can be made or purchased.

- What are the blueprint tolerances?
- What machine is being used?

The NC Setup Sheet

Once the process planning is finished and the program written, the programmer must communicate to the setup personnel in the shop what tools and fixtures are to be used in the NC program. This information is often placed on setup sheets such as those shown in Figures 3-1 and 3-2. The setup sheet should contain all necessary information to prepare for the job. Any special instructions to the setup personnel or machine operators should be communicated. Any special notes regarding tooling should also be included.

TOOLING FOR NUMERICAL CONTROL

Tooling is a vital consideration in efficient NC programming. This section is intended to show the prospective NC programmer various standard tooling options available. It is assumed the student has had exposure to cutting tools commonly used in a machine shop. A brief look through any beginning machining text will supply any necessary review of cutting tools.

Cutting Tool Materials

Cutting tools are available in three basic material types: high speed steel, tungsten carbide, and ceramic. The type of tool should be carefully chosen.

High speed steel (HSS) is one type of tool material. It has the following advantages over carbide:

- HSS costs less than carbide or ceramic tooling.
- HSS is less brittle and not as likely to break during interrupted cuts.
- The tools can be resharpened easily.

High speed steel tools have the following disadvantages:

- HSS does not hold up as well as carbide or ceramic at the high temperatures generated during machining.
- HSS does not cut hard materials well.

Tungsten carbide (known simply as carbide) is another material often used for cutting tools. Carbide tools come in one of three basic types. Solid carbide tools are made from a solid piece of carbide. Brazed carbide tools use a carbide cutting tip brazed on a steel shank. Inserted carbide tooling utilizes indexable inserts made of carbide which are held in steel tool holders. Tungsten carbide has the following advantages over high speed steel:

TAPE NUMBER: 1053

FIXTURE: 6 IN. MILL VISE

TABLE LAYOUT:

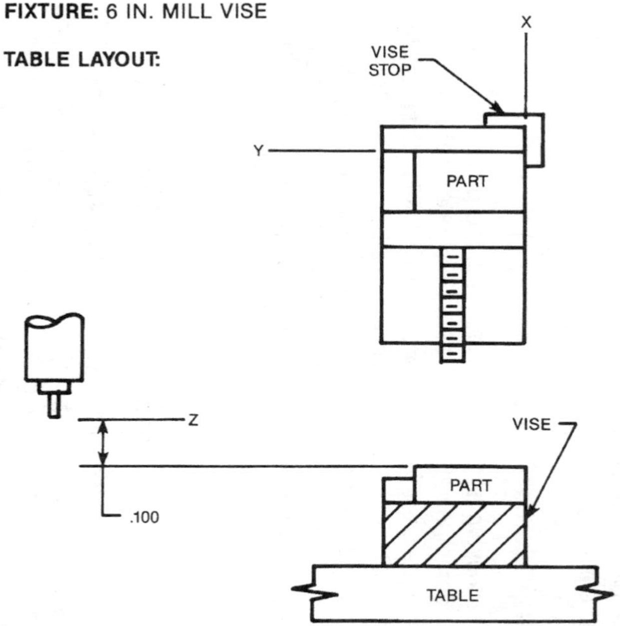

STA. NO.	CRO REG.	TOOL DESCRIPTION
1	—	3.0 DIA. INSERTED FACE MILL
		W/ .015 R GRADE 883 INSERTS
2	D12	.500 DIA. 4-FLUTE SOLID CARB. END MILL
3	—	NO. 4 × 90° C'DRILL
4	—	1/4 DRILL (.250 DIA.)
5	—	.262 DIA. BORING BAR

NOTES: DRILL POINT ANGLES TO BE 118° INCL.

NC SETUP SHEET FOR:

MACHINE: VERTICAL MACHINING CENTER

DRWN: WSS

PROG: WSS

DATE: 1-10-89

B/P REV: C

FIGURE 3-1
NC setup sheet for a CNC machining center

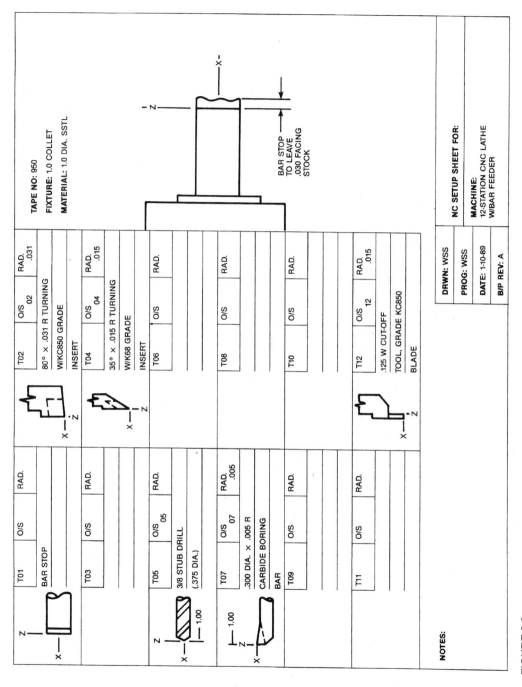

FIGURE 3-2
NC setup sheet for a CNC lathe

- Carbide holds up well at elevated temperatures.
- Carbide can cut hard materials well.
- Solid carbide tools absorb workpiece vibration and reduce the amount of "chatter" generated during machining.
- When inserted cutters are used, the inserts can be easily changed or indexed, rather than replacing the whole tool.

Carbide also has the following disadvantages:

- Carbide costs more than high speed steel.
- Carbide is more brittle than HSS, and has a tendency to chip during interrupted cuts.
- Carbide is harder to resharpen and requires diamond grinding wheels.

Ceramic tooling has made great advances in the past several years. While once very expensive, some ceramic inserts can now be purchased for less than the cost of carbide. Ceramic has the following advantages:

- Ceramic is sometimes less expensive than carbide when used in insert tooling.
- Ceramic will cut harder materials at a faster rate and has superior heat hardness.

Ceramic has the following disadvantages:

- Ceramic is more brittle than HSS or carbide.
- Ceramic must run within its given surface speed parameters. If run too slowly, the insert will break down quickly. Many machines do not have the spindle RPM range needed to use ceramics.

High speed steel is generally used on aluminum and other nonferrous alloys, while carbide is used on high silicon aluminums, steels, stainless steels, and exotic metals. Ceramic inserts are used on hard steels and exotic metals. Inserted carbide tooling is becoming the preferred tooling for many NC applications.

Some carbide inserts are coated with special substances, such as titanium nitride to improve the insert life. These coatings can increase tool life by up to 20 times when used in accordance with the manufacturer's recommended cutting speeds and feedrates.

TOOLING FOR HOLE OPERATIONS

There are four basic hole operations which are performed on NC machinery: drilling, reaming, boring, and tapping.

Drilling

Drills are available in different styles for different materials. Figure 3-3 shows a standard twist drill. Even with all the new tooling technology, twist drills remain one of the most common tools for making holes. Drills have a tendency to walk as they drill, resulting in a hole that is not truly straight. Center drills such as shown in Figure 3-4 are often used to predrill a pilot hole to help twist drills start straight. Drills also produce triangular-shaped holes.

If a hole tolerance is closer than .003 inch, a secondary hole operation should be used to size the hole, such as boring or reaming. Large holes are sometimes produced by spade drills (Figure 3-5). The flat blades allow good chip flow and economical replacement of the drill tip.

Drill point angle must be considered when selecting a drill. The harder the material to be cut, the greater the drill point angle needs to be to maintain satisfactory tool life. Mild steel is usually cut with a 118-degree included angle drill point. Stainless steels often use a 135-degree drill point.

Drills are available in different types. HSS drills are the most common, but brazed carbide and solid carbide are also used. Carbide drills have a tendency to chip when drilling holes. When drilling hard materials cobalt drills (HSS with cobalt added to the alloy) are used. Cobalt drills have greater heat hardness than HSS drills.

Special drills utilizing carbide inserts have been developed for NC applications (Figure 3-6). The economics of using these tools should be considered by the programmer when hard materials or high run quantities are involved.

FIGURE 3-3
Tapered shank twist drill *(Photo courtesy of Morse Cutting Tools Division)*

FIGURE 3-4
Center drill *(Photo courtesy of DoALL Manufacturing)*

FIGURE 3-5
Spade drill
*(Photo
courtesy of
DoALL
Manufacturing)*

**INSERT
POCKET A**
EDGES 1 & 2 USED
IN POCKET "A"

**INSERT
POCKET B**
EDGES 3 & 4 USED
IN POCKET "B"

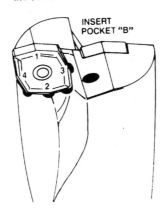

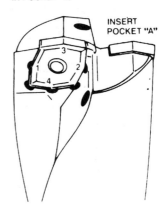

**4 CUTTING EDGES
FROM EACH INSERT**

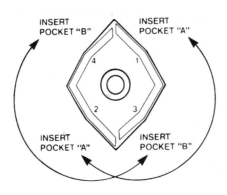

FIGURE 3-6
(Courtesy Carboloy Inc., A Seco Tools Company)

Reaming

Reaming is used to remove a small amount of metal from an existing hole as a finishing operation. Reaming is a precision operation which will hold a tolerance of + or −.0002 easily.

Reamers are made with two basic flute designs: straight fluted (Figure 3-7) and spiral fluted (Figure 3-8). Spiral fluted reamers produce better surface finishes than straight flutes, but are more difficult to resharpen. Reamers are available in three basic tool materials: high speed steel, brazed carbide, and solid carbide.

FIGURE 3-7
Straight flute chucking reamer *(Photo courtesy of Cleveland Twist Drill Company)*

FIGURE 3-8
Sprial flute chucking reamers *(Photo courtesy of DoALL Manufacturing)*

Boring

Boring removes metal from an existing hole with a single point boring bar. Boring heads are available in two designs: offset boring heads, in which the boring bar is a separate tool inserted into the head, and cartridge type. Cartridge boring heads use an adjustable insert in place of a boring bar.

Boring bars are available in the four material types: high speed steel, solid carbide, brazed carbide, and inserted carbide. Inserted carbide bars are used for large holes, whereas brazed and solid carbide bars are usually supplied in smaller sizes (up to $1/2$-inch diameter).

Tapping

Tapping is used to produce internally threaded holes. They are available in several flute designs. Standard machine screw taps (Figure 3-9) are widely used, especially when tapping blind holes. Spiral pointed taps (known as gun taps) are preferred for thru hole operations. These taps shoot the chips forward and out the bottom of the hole. High spiral taps (Figure 3-10) are used for soft stringy material such as aluminum.

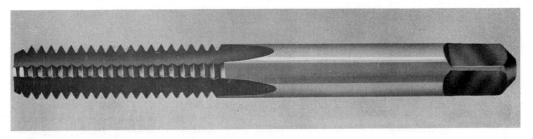

FIGURE 3-9
Machining screw tap *(Photo courtesy of Morse Cutting Tools Division)*

FIGURE 3-10
High spiral coated tap *(Photo courtesy of DoALL Manufacturing)*

A special milling cutter called a thread hob (Figure 3-11) is sometimes used to mill a thread in a workpiece. Thread hobs make use of an NC machine's helical interpolation capabilities. Helical interpolation is presented in Chapter 12.

FIGURE 3-11
Thread hob *(Photo courtesy of GET Valenite)*

MILLING CUTTERS

The greatest advances in tooling for NC have taken place in the area of inserted milling cutters. *Milling* allows the contouring capabilities of the NC machine to be used to efficiently perform operations that would require special tooling if done manually. Milling cutters can be placed in two basic categories: solid milling cutters and inserted milling cutters. They can be further classified as end mills and face mills.

End Mills

End mills are available in HSS and solid carbide from .032 inch to 2 inches in diameter in two or four flute. Inserted end mills are available from .500 inch to 3 inch diameters. Figure 3-12 shows a four flute HSS end mill. Figure 3-13 shows a two flute solid carbide end mill. Two flute cutters with their deeper gullets are well suited for roughing operations. Four flute end mills, however, are more rigid because of their thicker core. The programmer's experience will determine when to use a two or four flute cutter.

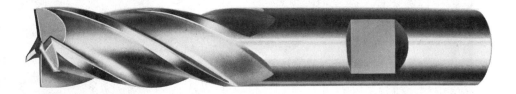

FIGURE 3-12
Single end, multiple flute end mill, standard length flutes *(Photo courtesy of Sharpaloy Division, Precision Industries, Inc.)*

FIGURE 3-13
Solic carbide, two-flute, end mill *(Photo courtesy of DoALL Manufacturing)*

Figures 3-14 and 3-15 illustrate two different types of inserted end mills. Inserted cutters are preferred for NC applications. Inserts are less expensive to replace than an entire tool. By indexing the inserts, four or six cutting edges can be used on one insert. When the insert is used up, it is thrown away rather than resharpened. Inserted cutters may also be used on many different types of workpiece materials by simply changing the inserts from one designed for aluminum, for example, to one designed for stainless steel.

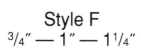

Style F
$^3/_4" — 1" — 1^1/_4"$

Style G
$1^1/_2" — 2"$

FIGURE 3-14
Inserted carbide end mills *(Photo courtesy of GTE Valenite)*

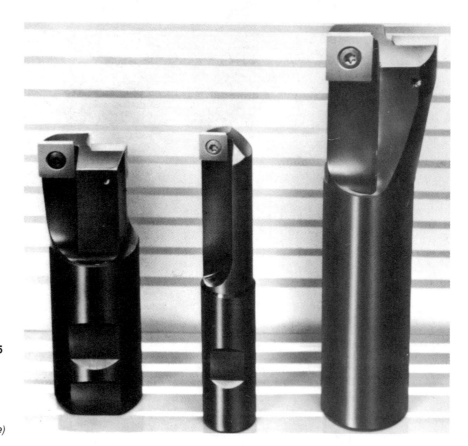

FIGURE 3-15
"Centerdex"
two-flute
inserted end
mills *(Photo
courtesy of
GTE Valenite)*

Figures 3-16 and 3-17 show two different styles of inserted ball end mills. Ball end mills are also available in HSS and solid carbide. Ball mills are used for three-, four-, or five-axis contouring work, where the Z-axis will be used. They are also used to produce a given radius on a part.

FIGURE 3-16
Ball nose end mills featuring round inserts *(Photo courtesy of GTE Valenite)*

FIGURE 3-17
Ball nose end mills featuring triangular inserts *(Photo courtesy of GTE Valenite)*

Figure 3-18 shows a special type of inserted end mill called a cyclo mill, designed by Valenite GTE. It uses a series of round inserts staggered in a helical pattern. This mill can remove large amounts of material at fairly high speeds. It is just one example of inserted tooling that is being developed for NC use.

FIGURE 3-18
"Cyclo Mill" special multi-inserted milling cutter *(Photo courtesy of GTE Valenite)*

Face Mills

Face mills differ from end mills in their major application. Face mills are designed to remove large amounts of material from the face of a workpiece. They are manufactured in HSS, brazed carbide, and inserted carbide types.

Face mills are available in sizes from 2 inches to over 8 inches in diameter. Inserted carbide is the most common type of facing tool. The costs of large brazed carbide and HSS mills limit their application to special situations.

Figure 3-19 shows a common type of inserted face mill. Figure 3-20 shows a large diameter face mill. Note the number of inserts used. In Figure

FIGURE 3-19
Carbide inserted face mill *(Photo courtesy of GTE Valenite)*

FIGURE 3-20
Large inserted face mill — note number of inserts on cutter *(Photo courtesy of GTE Valenite)*

3-21, a special type of mill cutter is shown. This cutter is called a plunge and profile cutter. It is designed to plunge into the material first and then begin the cutting path. This design is a cross between an end mill and a face mill.

FIGURE 3-21
Plunge and profile inserted milling cutter *(Photo courtesy of GTE Valenite)*

SPECIAL INSERTED CUTTERS

A number of special tools have been developed for uses with NC. The NC programmer is always confronted with new ideas to improve productivity. Prospective and experienced programmers should spend time looking at tooling catalogs to become acquainted with current tooling developments.

Figures 3-22 through 3-24 illustrate some of the current tooling ideas developed specifically for NC applications.

FIGURE 3-22
Special small diameter inserted end mill *(Photo courtesy of GTE Valenite)*

Carbide Inserts and Their Selection

Carbide inserts are manufactured in a variety of types and grades. The type of insert describes the shape of the insert. Some common shapes include triangular, 80-degree diamond, 55-degree diamond, and round. The grade of insert refers to the hardness of the insert and application for which it was developed. The NC programmer must be aware of the type and grade of insert available when making tooling selections. Each type of insert is iden-

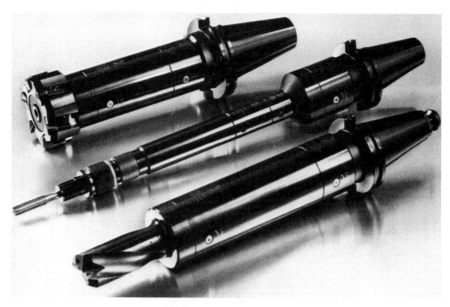

FIGURE 3-23
Special inserted tooling for use with NC. From top to bottom: an inserted milling cutter with interchangeable tooling extensions, a machine tap in a tap holder with interchangeable tooling extensions, and an inserted drill mounted in a holder with interchangeable extensions *(Photo courtesy of GTE Valenite)*

tified by a designation code. The identification system used on an insert will vary depending on the manufacturer. Figure 3-25 shows one such system.

Figure 3-26 illustrates some of the carbide grades available and their applications. Each grade of carbide is designated by an ANSI "C" number designation from C-1 to C-8. In addition, each grade of carbide has also been classified by the ISO. The ISO designation uses a "K" or "P" number, depending on insert hardness. In the United States, the ANSI system is generally used. In other countries, the ISO system is followed. Manufacturers develop their own grade system based on the ANSI or ISO rating. A C-2 ANSI grade insert would be called CQ2 by RWT Corporation, K1 by Kennametal Inc., and VC-1 by Valenite GTE Inc. It is necessary therefore, for the programmer to consult the individual manufacturer's catalog to arrive at the proper grade number.

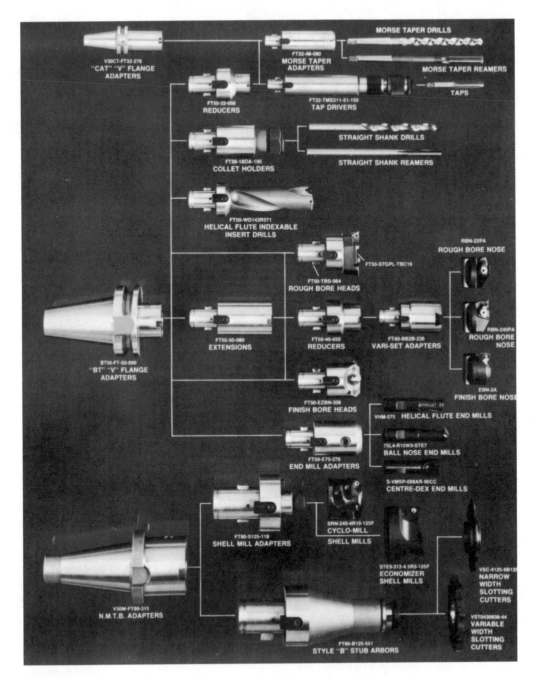

FIGURE 3-24
An NC tooling system featuring tool adapters, interchangeable extensions, tool bodies, boring heads, and arbors *(Photo courtesy of GTE Valenite)*

Identification System

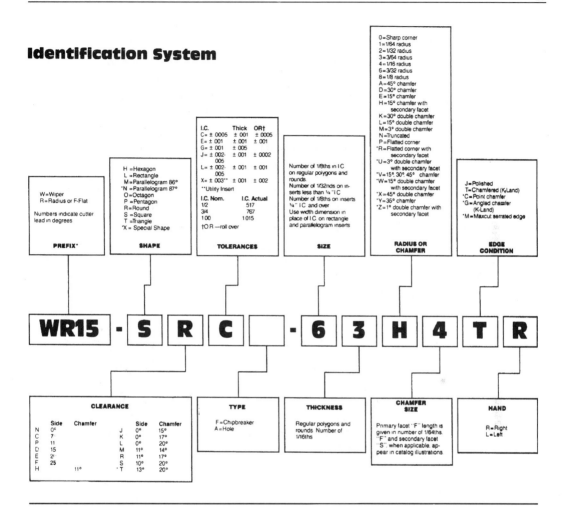

FIGURE 3-25
Carbide insert identification system *(Courtesy of Carboloy Inc., A Seco Tools Company)*

Lathe Tooling

Carbide inserted tools dominate tool selection for CNC turning applications. One of the more popular insert styles is the diamond insert. Figure 3-27 illustrates a series of inserted turning tools. Figure 3-28 shows a number of boring bars and toolholders.

INSERT GRADE APPLICATION CHART

Cast iron and nonferrous materials	Alloy and tool steels Stainless steels
C-1: Roughing	C-5: Roughing
C-2: General Purpose	C-6: General Purpose
C-3: Finishing	C-7: Finishing
C-4: Precision Finishing	C-8: Precision Finishing

MANUFACTURER'S GRADE DESIGNATION

ANSI Class	ISO Class	Carboloy	Iscar	Kennametal	Sandvik	Valenite
C-8	P-01 P-05	210	IC-80t	K7H	F02	VC-8
C-7	P-10 P-25	350	IC-70	K45	S1P	VC-7
C-6	P-25 P-35	370	IC-50	KC850	S4	VC-55
C-5	P-40 P-50	518	IC-54	—	S35	VC-5
C-4	K-01 K-05	999	IC-4	K11	—	VC-4
C-3	K-10 K-15	905	IC-20	K68	H10	VC-3
C-2	K-20 K-25	883	IC-2	K6	H20	VC-2
C-1	K-30 K-20	820	IC-28	K1	H	VC-1

Note: Most manufacturers produce more than one grade per insert class. Consult the manufacturer's catalog for a complete listing.

FIGURE 3-26
Carbide insert grades

A PROCESSING EXAMPLE

In a large company, the formal processing (determining the machine routing through the shop) is done by a processing engineer. The process is then sent to the programming department where the tooling concepts and machining strategy is done. In a small company, the NC programmer does both the processing and programming. It is important that the department processing a job work closely with the programming department to efficiently and economically produce a manufacturing process.

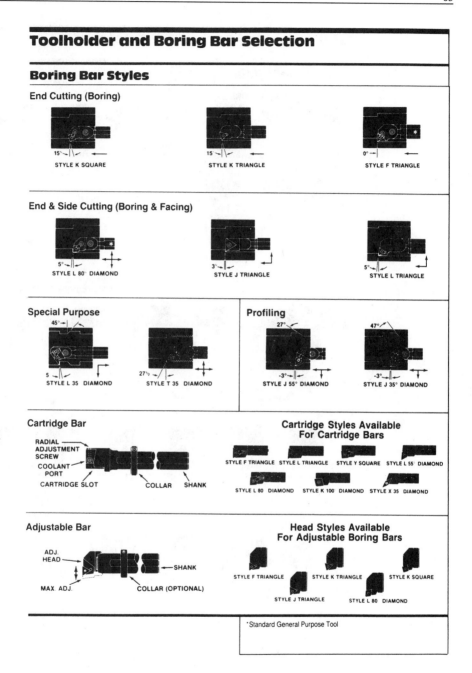

FIGURE 3-27

Toolholder and boring bar sections — boring bar styles *(Courtesy of Carboloy Inc., A Seco Tools Company)*

Toolholder and Boring Bar Selection

Toolholder Styles

Side Cutting Tools (Turning)

STYLE D SQUARE STYLE R SQUARE STYLE R TRIANGLE STYLE W TRIANGLE STYLE G TRIANGLE STYLE A TRIANGLE STYLE E TRIANGLE

STYLE M
100 —80 /100

STYLE R
100 —80 /100

STYLE G
80 —80 /100

End Cutting Tools (Facing)

  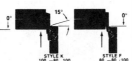

STYLE K SQUARE   STYLE F TRIANGLE   STYLE K
100 —80 100

STYLE F
80 —80 /100

Side & End Cutting Tools (Turning & Facing)

STYLE G ROUND   STYLE S SQUARE   STYLE L
80 —80 /100

Profiling Tools

STYLE L TRIANGLE   STYLE J TRIANGLE   STYLE J 55 DIAMOND   STYLE P 55 DIAMOND   STYLE J 35 DIAMOND   STYLE L 35 DIAMOND

Special Purpose Tools

A   G O
TEE LOCK

STYLE C
TRIANGLE

STYLE T
35° DIAMOND

O   G
ROUND

' Standard General Purpose Tool

FIGURE 3-28
Toolholder and boring bar selection — toolholder styles *(Courtesy of Carboloy Inc., A Seco Tools Company)*

Figure 3-29 is a part that is to be machined from an aluminum casting. The casting has .250 dia. of stock to be removed from the 4.000 and 3.000 diameters. The center of the casting was cored to 1.000 inch, and the 1.00 height was cast at 1.250. After consultation with the NC programming department, the process illustrated in Figure 3-30 was developed. The 4.000

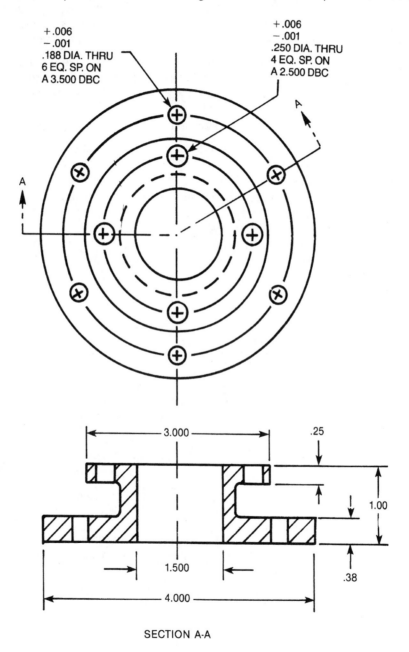

SECTION A-A

FIGURE 3-29
Part drawing

inch diameter and .38 dimension are to be done on a conventional lathe. The part will then be routed to a vertical spindle CNC machining center where the balance of the part is to be completed.

The fixture concept to hold the part was developed by the NC programmer. The concept drawing is shown in Figure 3-31. The part will be nested in the 4.0015 diameter fixture bore. It will be clamped with four swiveling clamps, available as a purchased item from a tooling component supplier. This fixture design was based on the following factors.

- The 4.000 diameter and .38 dimensions were completed in the previous operation, making this feature the logical choice for locating the part.
- The run quantity is only 200 parts. The fixture design is simple, making it economical to build.
- The design is easy to load.

The sequence of the machining operation at the machining center was planned as follows.

- Face the 1.000 and .25 dimensions using a $3^{1}/_{4}$ carbide inserted face mill.
- Center drill the .188 and .250 diameter holes. A 90-degree center drill was chosen. The 90 degree-chamfer will provide an edge break at the drilled hole, reducing the amount of deburr time.
- Drill the .188 diameter holes using a $^{3}/_{16}$ drill. Since drills almost always drill .001 or more oversize, the hole will be comfortably within tolerance.

MANUFACTURING PROCESS

Part Number: Adapter
Run Quantity: 200

Job Number: 000-000-001
Material: Alum. Casting.

OPERATION NUMBER	OPERATION CODE	DESCRIPTION OF OPERATION
010	issue	Issue 356 alum. castings
020	manual lathes	Chuck on 3.250 as cast dia. • Turn 4.000 ± .010 b/p dim to 4.000 ± .001 dia. (tooling dimension). • Face .38 b/p dim.
030	vert. mach. center	Locate parts in fixture NCF-000-100 • Drill .188 + .006 − .001 dia. thru 6 plcs. • Drill .250 + .006 − .001 dia. thru 4 plcs. • Bore 1.500 ± .010 dia. thru 1 plc. • Mill the 3.000 ± .010 dia., hole the 1.000 and .25 dims.
040	burr	• Deburr parts as required.
050	insp	• Inspect parts for b/p conformance.

FIGURE 3-30
Manufacturing process for part shown in Figure 3-29

- Drill the .250 diameter hole using a ¼ drill.
- Mill the 3.000 diameter using a 1¼ diameter inserted helical end mill. The end mill has inserts up the sides of the insert, allowing side cutting up to 2.00 deep.
- Using the same end mill, mill the 1.500 diameter bore.

The setup sheet for the NC operation is shown in Figure 3-32.

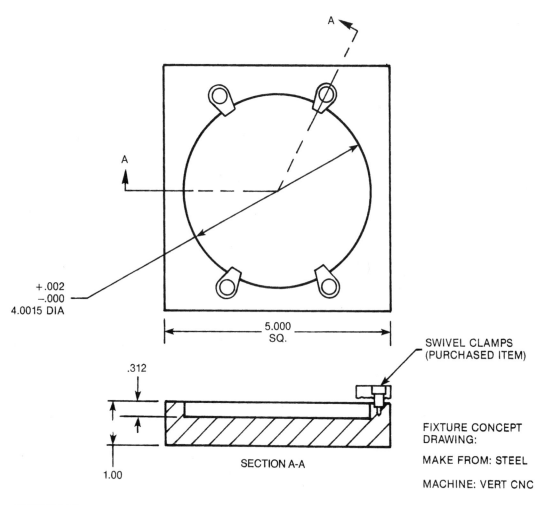

FIGURE 3-31
Fixture concept

TAPE NUMBER: 1000

FIXTURE: NCF-000-100

TABLE LAYOUT:

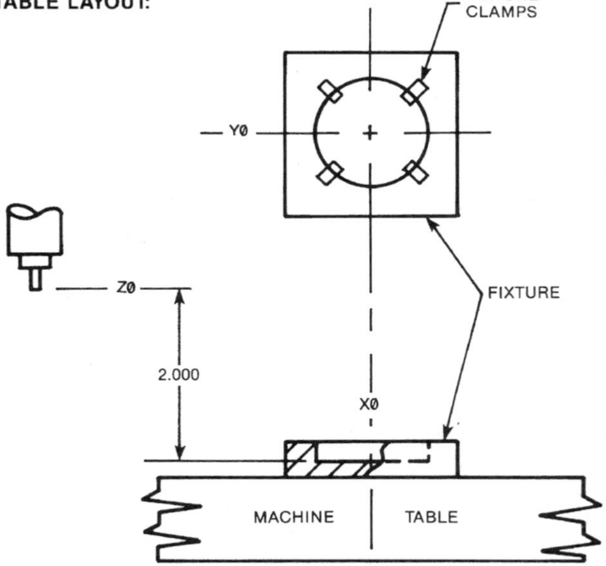

FIXTURE CLAMPS

FIXTURE

X0

Y0

Z0

2.000

MACHINE

TABLE

STA. NO.	CRO REG.	TOOL DESCRIPTION
1	D11	3 1/4 INSERTED CARBIDE FACE MILL
2		NO. 4 × 90° C'DRILL
3		3/16 DRILL (.1875 DIA.)
4		1/4 DRILL (.250 DIA.)
5	D15	1 1/4 INSERTED CARBIDE HELICAL END MILL

NOTES: TOOL NO. 2 REQUIRES 1.125 MIN EFF. LENGTH

SETUP SHEET FOR
CNC MACHINING CENTER

MACHINE: UNIVERSAL VERT. MACH. CENTER

DRWN: WSS

PROG: WSS

DATE: 3-4-89

B/P REV: A

OPER. NO: 030

FIGURE 3-32
NC setup sheet for CNC machining center

SPEED AND FEEDS

The efficiency and life of a cutting tool depend upon the cutting speed and the feedrate at which it is run.

Cutting Speed

The *cutting speed* is the edge or circumferential speed of a tool. In a machining center or milling machine application, the cutting speed refers to the edge speed of the rotating cutter. In a turning center or lathe application, the cutting speed refers to the edge speed of the rotating workpiece. Cutting speed (CS) is expressed in surface feet per minute (SFM). It is the number of feet a given point on a rotating part or cutter moves in one minute.

Proper cutting speed varies from material to material. Generally, the softer the material, the higher the cutting speed. Recommended cutting speeds for various materials can be found in tables contained in machinists' handbooks, and tooling manufacturers' catalogs. Appendix 6 of this text contains one such chart.

It should be understood that cutting speed and spindle RPM are two different things. A .250-inch diameter drill turning at 1200 RPM has a cutting speed of approximately 75 surface feet per minute. A .500-inch diameter drill turning 1200 RPM has a cutting speed of approximately 150 SFM. The spindle RPM necessary to achieve a given cutting speed can be calculated by the formula:

$$\text{RPM} = \frac{\text{CS} \times 12}{\text{D} \times \pi}$$

Where: CS = cutting speed in surface feet per minute
        D = diameter in inches of the tool (workpiece diameter for lathes)
        π = 3.1416

The cutting speed of a particular tool can be determined from the RPM, using the formula:

$$\text{CS} = \frac{\text{D} \times \pi \times \text{RPM}}{12}$$

On the shop floor, the formulas are often simplified. The following formulas will yield results similar to the formulas just given.

$$\text{RPM} = \frac{\text{CS} \times 4}{\text{D}}$$

$$\text{CS} = \frac{\text{RPM} \times \text{D}}{4}$$

For turning applications, the diameter of the workpiece, rather than the tool diameter, is used to determine the cutting speed and spindle speed. For milling applications, the diameter of the tool is used.

Feedrates

Feedrate is the velocity at which a tool is fed into a workpiece. Feedrates are expressed two ways: inches per minute of spindle travel and inches per revolution of the spindle. For milling applications, feedrates are generally given in inches per minute (IPM). For turning they are expressed most often in inches per revolution (IPR).

Feedrates are critical to the effectiveness of a job. Too heavy a feedrate will result in premature dulling and burning of tools. Feedrates which are too light will result in tools chipping. This chipping will rapidly lead to tool burning and breakage.

Turning Feedrates

The vast majority of turning tools used with NC are inserted tools. The feedrates used vary with material type and insert type. Tables found in manufacturers' catalogs and machining data handbooks are the best sources for turning feedrates. It must be noted that conditions such as part geometry, machine rigidity, and rigidity of the setup will affect both speeds and feedrates. The values given in the tables are starting points. The actual speed and feedrate used during the run will ultimately be determined during the job setup, when the first piece is run.

Drilling Feedrates

Drilling feedrates are dependent on the drill diameter. Tables in machinists' handbooks will list recommended feedrates in IPR forgiven diameters of a given tool material. For example, HSS drills from $1/8$ to $1/4$ inch use feedrates of .002 to .004 IPR. Drills from $1/4$ to $1/2$ inch use feedrates of .004 to .007 IPR. Drills from $1/2$ to 1 inch use feedrates from .007 to .015 IPR. The final feedrate used will depend upon these factors.

For machining center use, the feedrates given in the tables will have to be converted to IPM values. To accomplish this, the following formula is used:

$$IPM = RPM \times IPR$$

Where: IPM = the required feedrate expressed in inches per minute
RPM = the programmed spindle speed in revolutions per minute
IPR = the drill feedrate to be used expressed in inches per revolution

Milling Feedrates

Feeds used in milling depend not only on the spindle RPM, but also on the number of teeth on the cutter. The milling feedrate is calculated to produce a desired chip load on each tooth of the cutter. In end milling, for example, chip load should be .002 to .006 inch per tooth. The recommended chip loads for various mill cutters are given in machinists' handbooks. For inserted cutters, the insert manufacturer's catalog will list recommended chip loads for a given insert. To calculate the feedrate for a mill cut, the following formula is used:

$$F = R \times T \times RPM$$

Where: F = the milling feedrate expressed in inches per minute
          R = the chip load per tooth
          T = the number of teeth on the cutter
        RPM = the spindle speed in revolutions per minute.

Milling feedrates are also affected by machine and setup rigidity, and by part geometry.

In the case of inserted milling cutters, there is another factor which affects feedrates: chip thickness. This is not the chip load on the tooth, but the actual thickness of the chip produced at a given feedrate. Chip thickness will vary some with the geometry of the cutter (positive rake, negative rake, neutral rake), but should be maintained in the range of .004 to .008 inch. Chip thickness less than or greater than these values will place either too little or too great a pressure on the insert for efficient machining. Once a feedrate has been calculated, the chip thickness it produces should be derived. If chip thickness is out of the recommended range, the feedrate should be adjusted to bring it to acceptable limits.

Chip thickness can be calculated by the following formula:

$$CT = \sqrt{\frac{W}{D}} \times R$$

Where: CT = the chip thickness
         W = the width of the cut
         D = the diameter of the cutter
         R = the feed per tooth

If the chip thickness is found to be too small, this modification of the preceding formula can be used to determine an acceptable feedrate:

$$f = \sqrt{\frac{D}{W}} \times CT$$

Where: f = the feed per tooth being calculated
          D = the diameter of the cutter
         CT = the desired chip thickness

This new calculated value of the feed per tooth can then be substituted back into the feedrate formula, and a new feedrate calculated.

Speed and Feed Example

An aluminum workpiece is to be milled using a carbide inserted mill cutter. The cutter is 1.750 diameter × 4 flute. What would be the appropriate spindle RPM and milling feedrate for the workpiece?

An appropriate cutting speed (SFM) for aluminum is 1000 surface feet per minute. Using this value in the spindle speed formula with a cutter diameter of 1.75:

$$RPM = \frac{1000 \times 12}{1.75 \times 3.1416}$$

$$RPM = \frac{12000}{5.4978}$$

$$RPM = 2,183$$

The feedrate can now be determined using the feedrate formula. The machinist handbook's tables give a recommended chip load of .002 to .006 inch. A value of .004 per tooth is selected. Using these values in the feedrate formula, the feedrate is calculated.

$$F = 2,183 \times 4 \times .004$$
$$F = 34.91 \text{ inches per minute}$$

The chip thickness is then calculated to insure the inserts will not break down prematurely. For this example, it will be assumed the width of the cut to be taken is 1.000 inch wide. Using this value, the chip thickness is determined as follows:

$$CT = \sqrt{\frac{1.000}{1.750}} \times .004$$

$$CT = .755 \times .004$$

$$CT = .00302$$

The chip thickness is less than the recommended minimum of .004. The feed per tooth is therefore calculated as using the feed per tooth formula. A chip thickness of .008 is used.

$$f = \sqrt{\frac{1.75}{1.000}} \times .008$$

$$f = 1.3229 \times .008$$

$$f = .010$$

The new value for the chip load per tooth is substituted in the feedrate formula, and the feedrate recalculated.

$$F = 2183 \times 4 \times .010$$
$$F = 87.32 \text{ inches per minute}$$

The 2183 RPM spindle speed and 87.32 inches per minute feedrate are "book value" rates. They will have to be adjusted up or down depending on the machine, fixture, tool, and workpiece rigidity.

SUMMARY

The important concepts presented in this chapter are:

- Process planning is the term used to describe the steps the programmer uses to develop and implement a part programming.
- The steps in process planning are: determine the machine, determine the workholding, determine the machining strategy, select the tools to be used.
- Tool selection is important to the efficiency of the NC program.
- Cutting tools for NC are made in high speed steel, tungsten carbide, and ceramic.
- Inserted cutters are the preferred tools for NC use.
- Inserts are manufactured in different grades with different applications intended.
- Cutting speed is the edge speed of the tool; it is a function of the spindle RPM and the tool diameter.
- Feedrates that are too heavy will result in excess tool wear and premature tool failure.
- Feedrates that are too light will result in chipping of tools and premature tool failure.
- When calculating milling feedrates, chip thickness must be considered.

VOCABULARY INTRODUCED IN THIS CHAPTER

Chip thickness
Cutting speed (CS)
Feedrate
High speed steel (HSS)
Methodizing
Process planning
NC Setup Sheet
Tungsten carbide

REVIEW
QUESTIONS

1. What is process planning?
2. List three factors that influence NC machine selection.
3. List three factors used by the programmer in deciding how to hold a workpiece.
4. List two factors that influence the development of machining strategy.
5. List three factors that determine tool selection.
6. What is an NC setup sheet?
7. What are the three basic cutting tool materials?
8. Give one advantage of HSS tools? Give one disadvantage.
9. Give one advantage of carbide tools. Give one disadvantage.
10. On what types of materials are ceramic inserts used?
11. What types of tools are preferred for NC use? Why?
12. What are the four common hole operations performed on NC machines?
13. Why are drilled holes not perfectly straight?
14. Is drill point angle important? Why?
15. Which reamer flute design produces better finishes: straight or spiral?
16. What are the two types of boring head?
17. What is the difference between a tap and a hob?
18. What are the two categories of milling cutters?
19. Into what further classifications can the categories in #18 be put?
20. Why should a programmer be familiar with insert grades?
21. What is cutting speed? Is it the same as the spindle RPM?
22. What happens when a feedrate is too light? too heavy?
23. What two ways are feedrates expressed?
24. How is a milling feedrate calculated?
25. Why should chip thickness be checked after a feedrate is determined?

CHAPTER 4

Tool Changing and Tool Registers

OBJECTIVES Upon completion of this chapter, you will be able to:

- Explain why the speed, repeatability, and accuracy of tool changing are important factors in numerical control.
- Name the two types of tool changes.
- Explain why quick-change tooling is used on NC mills.
- Explain how tooling is used in automatic tool change functions.
- Name the five types of automatic tool changers and briefly describe the operation of each.
- Describe the two basic methods of tool storage.
- Explain what tool registers are and what they are used for.
- Describe what tool offset length is and how it is determined.
- Explain how tool offsets may be entered by the operator during setup and how the programmer allows for this.

   This chapter deals with CNC tool changing and tool registers. A good general understanding of these subjects is required for three-axis CNC programming.

TOOL CHANGES

   There are two types of tool changes: *manual* and *automatic*. When referring to CNC mills, tool changing is understood to be manual unless otherwise stated. A machining center, on the other hand, incorporates automatic tool change (ATC). It is the tool-changing capability that separates the CNC machining center from CNC milling machines. Machining centers, like milling machines, have the capability to do numerous machining operations (drilling, tapping, spotfacing, and milling, among others). This is opposed to a machine capable of a single function only, such as an NC drilling machine. Figure 4-1 illustrates a CNC milling machine; Figures 4-2 and 4-3 illustrate CNC machining centers. Notice

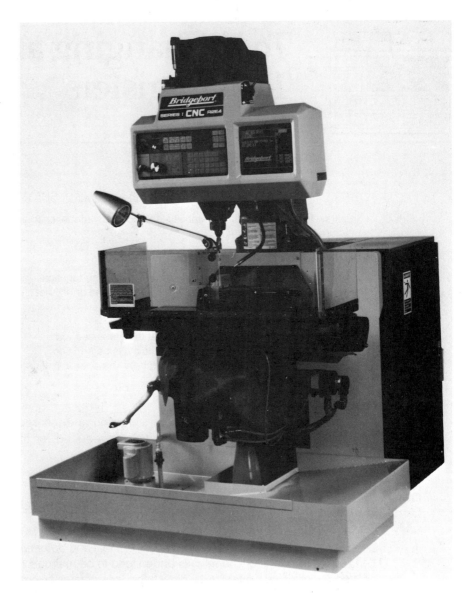

FIGURE 4-1
A vertical spindle CNC milling machine. Note the quick change tooling system installed in the spindle. *(Photo courtesy of Bridgeport Machines Division of Textron Inc.)*

the presence of the tool changer on the machining centers. Also note that a machining center may have either a horizontal or vertical spindle just like a milling machine.

FIGURE 4-2
A horizontal CNC machining center employing automatic tool change. Note the pivot insertion tool changer on the side. Tools are stored in a matrix magazine. Safety guards
have been removed for clarity. *(Photo courtesy of Cincinnati Milacron)*

Tooling for Manual Tool Change

What is to be gained by the speed with which a CNC machine can position itself for hole drilling if the tool changes are so lengthy as to cancel the
time and accuracy gained by using numerical control? Tool changing greatly
influences the efficiency of numerical control, so tool changes should take
place as quickly as is safely possible. The tool must not only be (1) accurately
located in the spindle to assure proper machining of the workpiece, but (2) the
tool must be located as accurately as possible in the same location and (3) in
the same relationship to the workpiece each time it is inserted in the spindle.
This is known as *repeatability* of a tool—the ability to locate, or repeat, its
position in the spindle each time it is used.

Numerical control mills (manual tool change) usually are supplied with or
have had added to them some type of quick-change tooling system to accomplish this task.

FIGURE 4-3
A vertical spindle CNC machining center *(Photo courtesy of Cincinnati Milacron)*

Most small vertical turret mills are manufactured with what is known as an R-8 spindle taper, which will accept R-8 collets. Figure 4-4 depicts an R-8 spindle and collet. The CNC milling machine in Figure 4-1 has an R-8 spindle employing a quick-change tool-changing system. The R-8 collet is a standard collet on Bridgeport vertical mills. Since most vertical turret mills are spinoffs of this design, the R-8 spindle has become pseudo-standard on these machines.

R-8 collets and R-8 tool holders require the use of a drawbar. For CNC use, either an automatically tightening drawbar is supplied with the machine, or a quick-change tool system is added.

The quick-change tool system consists of a quick release chuck (which is held in the machine spindle) and a set of tool holders that hold the individual tools needed for a particular part program. The chuck is a separate tool-holding mechanism that stays in the spindle. During a tool change, the tool holder is removed from the chuck (sometimes called the tool changer), and a tool holder containing the next required tool is installed in its place. The tools

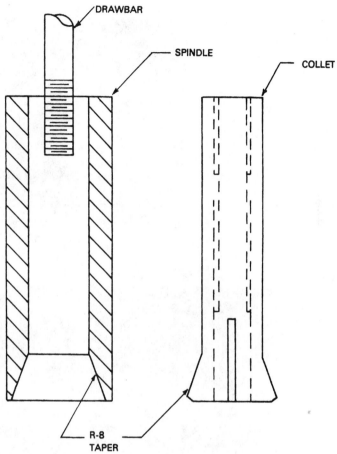

FIGURE 4-4
R-8 spindle and collet

placed in the tool holders are securely held by means of set screws. Many varieties of these quick-change tool systems are available on the market. Figure 4-5 illustrates a quick-change tooling system in action.

Larger vertical mills and most horizontal mills use another type of spindle taper, called the American Standard Milling Machine Taper (Figure 4-6). Like the R-8 taper, this taper requires the use of a drawbar. If no automatic drawbar is supplied with a machine, a quick-change tooling system is added to the machine to improve tool changing.

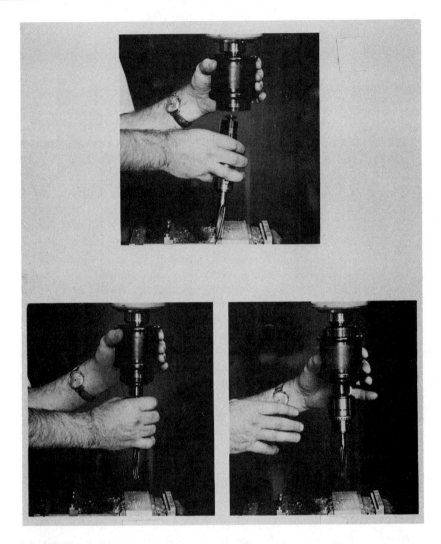

FIGURE 4-5
A quick change tooling system used for manual tool change *(Photo courtesy of Immotion Quick Change Tool Systems)*

Tooling for Automatic Tool Change

When automatic tool change is used, the requirements for speed and repeatability are even more critical. The machine's tool changer cannot think and correct for misalignment of tooling or tool setup errors like a human being. The tool changer will faithfully carry out its tool-changing cycle and nothing else (that's all it was programmed to do). Tooling used with a tool changer,

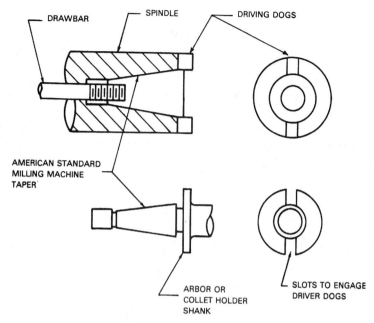

FIGURE 4-6
American Standard Milling Machine Taper used on spindle and arbor (or collet holder shank)

therefore, must be (1) easy to center in the spindle; (2) easy for the tool changer to grab; and (3) have some means providing for the safe disengagement of the tool changer from the tool once secured in the spindle. Figure 4-7 depicts a common type of tool holder used with ATC (automatic tool change).

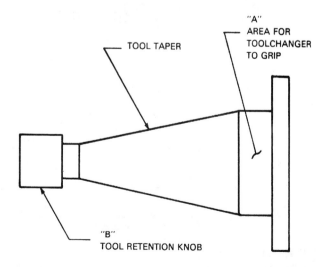

FIGURE 4-7
Typical toolholder used with ATC

The tool changer grips the tool at point A in Figure 4-7 and places the tool in position, aligned with the spindle. The tool changer will then insert the tool into the spindle. In some cases, insertion of the tool is accomplished by the spindle descending over the tool. As the tool engages the spindle, a split bushing in the spindle will close on the *tool retention knob* (point B in Figure 4-7): This split bushing holds the tool so that the tool changer can release its grip on the tool. The tool is then drawn completely up into the spindle and tightened. Using this procedure insures proper alignment of the tool with the spindle and prevents damage from occurring to the spindle or tool holder taper. Figure 4-8 shows tool insertion using a split bushing.

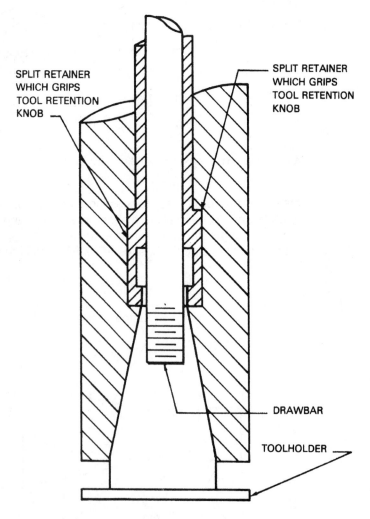

FIGURE 4-8
Split bushing closes over the retention knob to secure the tool as it is drawn into the spindle

Another insertion method can be used with a different type of tool holder (Figure 4-9). Here, the tool changer grips the tool in slot A. After the tool is inserted into the spindle, the tool changer moves toward the spindle as the tool is drawn up into the spindle. When the tool is secured in the spindle, the tool changer slides off the tool holder from the side.

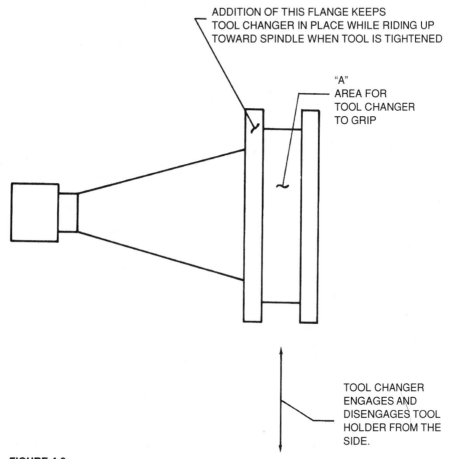

ADDITION OF THIS FLANGE KEEPS
TOOL CHANGER IN PLACE WHILE RIDING UP
TOWARD SPINDLE WHEN TOOL IS TIGHTENED

"A"
AREA FOR
TOOL CHANGER
TO GRIP

TOOL CHANGER
ENGAGES AND
DISENGAGES TOOL
HOLDER FROM THE
SIDE.

FIGURE 4-9
Tool changer moves in from the side to grip the toolholder in area A while the tool is secured in the spindle

AUTOMATIC TOOL CHANGERS

Automatic tool changers, while varied, are made in five basic types: turret head, 180-degree rotation, pivot insertion, multi-axis, and spindle direct. Tools used in automatic tool change are secured in tool holders designed for that purpose. These tool holders are installed directly in the spindle at each tool change by the tool changer. An assortment of tools and tool holders used with CNC machining centers is shown in Figure 4-10.

FIGURE 4-10
An assortment of tools and toolholders used with CNC machining center *(Photo courtesy of Command Corporation International)*

Turret Head

Tool changing accomplished through the use of a turret head is perhaps the oldest form of automatic tool change. A *turret head* is a number of spindles linked to the same milling machine head, as depicted in Figure 4-11. The tools are placed in the spindles prior to running the program. When another tool is needed, the head *indexes* (moves) to the desired position.

The main disadvantage of this system is the limited number of tool spindles available. In order to use a greater number of tools than available spindles, the operator must remove tools that have already been used and insert those called for later in the program. While other tool-changing methods require less machine operator attention once the program is running, no tool removal is actually performed during the tool change. This results in a very quick tool change. Turret heads are still being used today on certain types of NC machinery such as drilling machines.

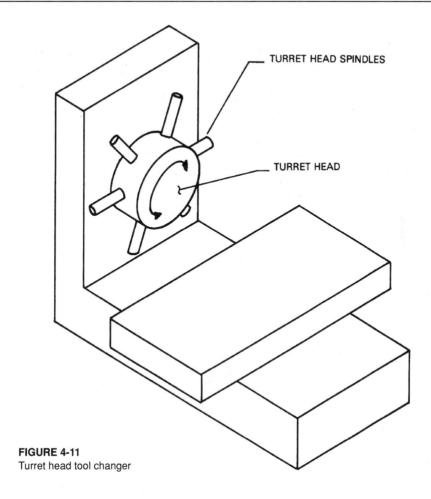

TURRET HEAD SPINDLES

TURRET HEAD

FIGURE 4-11
Turret head tool changer

180-Degree Rotation

The simplest of the true tool-changing mechanisms is the *180-degree rotation* tool changer (see Figure 4-12). Upon receiving a tool change command, the machine control unit sends the spindle to its fixed tool change coordinates. At the same time, the tool magazine is indexed to the proper position. The tool changer then rotates and engages both the tool in the spindle and the tool in the magazine at the same time. The drawbar is removed from the tool in the spindle, and the tool changer removes both tools from their respective places. The tool changer then rotates 180 degrees and swaps the tool that was in the spindle with the one that was in the magazine. While the tool changer is rotating, the magazine repositions itself to accept the old tool that was removed from the spindle. The tool changer then installs the new tool in the spindle and the old tool in the magazine. Finally, the tool changer rotates back to its "parked" position where it remains until needed. The tool change is thus complete and the program continues.

The principal advantage of this type of changer is its simplicity. The amount of motion involved is minimal and tool changes are fast. The principal disadvantage is that the tools must be stored in a plane parallel to the spindle. The chances of chips and coolant getting on the toolholders are greatly increased compared to those in *side-* or *back-mounted magazines.* Extra protection for the tools must, therefore, be provided. Chips on the toolholder taper will also cause an inaccurate tool change, possibly damaging both the toolholder and

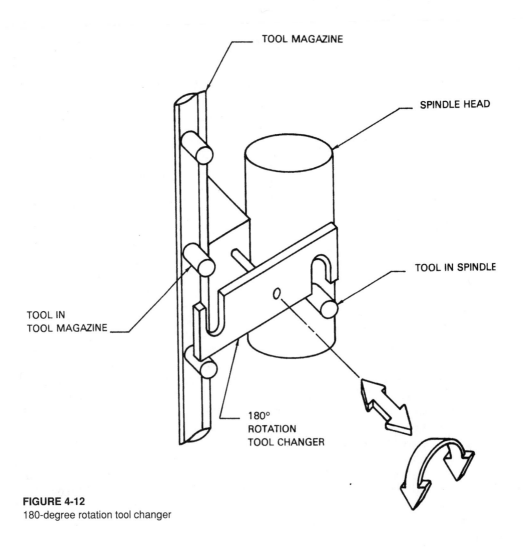

FIGURE 4-12
180-degree rotation tool changer

the spindle. Some machining centers employ a transfer arm that allows the tool magazine to be stored on the side of the machine. When the tool change command is issued, the transfer arm removes the tool from the magazine and pivots to the front of the machine, positioning the tool to be engaged by the tool changer. The 180-degree rotation tool changer may be used on either horizontal or vertical spindle machines.

Pivot Insertion

An adaptation of the 180-degree rotation tool changer is the *pivot insertion* tool changer (one of the most popular types in use). A pivot insertion system combines the functions of the tool changer and transfer arm. The operation of a pivot insertion tool changer is depicted in Figure 4-13. Figure 4-14 shows a pivot insertion tool changer on a horizontal machining center. This tool changer has the same physical design as that of the 180-degree rotation tool changer.

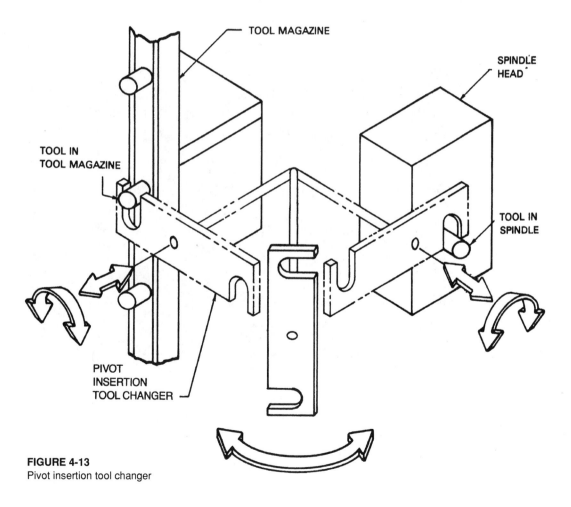

FIGURE 4-13
Pivot insertion tool changer

When a tool change command is given, the spindle is sent to the tool change location, and the tool magazine is rotated to the proper location for the tool changer to remove the new tool from its slot. The tool changer rotates and removes the new tool from the magazine, which is located on the side of the machine. The tool changer then pivots around to the front of the machine where it engages and removes the tool from the spindle, rotates 180 degrees, and inserts the new tool in the spindle. During this time, the tool magazine has indexed to the proper position to receive the old tool. The tool changer then pivots around to the side of the machine and places the old tool in its slot in the tool magazine. Finally, the tool changer "parks," and the NC program continues.

The main advantage of this system is that the tools may be stored on the side of the machine away from potentially damaging chips. Its disadvantage as compared to the 180-degree rotation tool changer is that pivot insertion requires more motion and therefore results in a more time-consuming tool change.

FIGURE 4-14
A pivot insertion tool changer on a horizontal machining center using twin matrix tool storage magazines. Guards have been removed for clarity. *(Photo courtesy of Cincinnati Milacron)*

Multi-Axis

Multi-axis tool change operation is depicted in Figure 4-15. This type of tool changer can be used with either side-mounted or back-mounted tool magazines. Its design lends itself very well to use with vertical spindle machining centers. When given a tool change command, the tool changer moves from its "parked" position, grabs the tool that is in the spindle, and removes it. The tool changer then swings (or sweeps) back to the tool magazine and places the old tool into the magazine. The changer then removes the desired tool from the magazine, swings around to the spindle again, and installs the tool in the spindle. Finally, the tool changer returns to "park," and the tool change is completed.

The main advantage of this system is the placement of the tool magazine on the back or side of the machine, where maximum protection can be afforded to the tools. Its disadvantage is the amount of tool handling and motion that must be employed. Today, multi-axis tool changers are giving way to other tool-changing mechanisms such as the 180-degree rotation, and, on vertical spindle machining centers, the spindle direct tool changer.

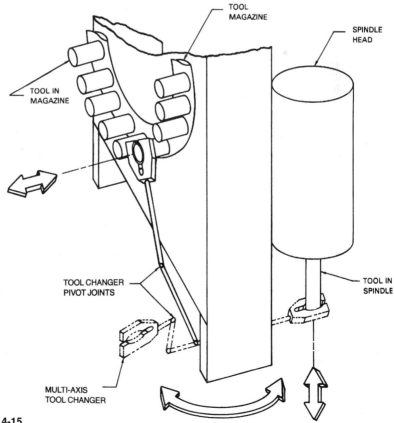

FIGURE 4-15
Multi-axis tool changer

Spindle Direct

Spindle direct tool changing differs from other types of tool changing in that the tool magazine (carousel) moves directly to the machine spindle or vice versa. Figure 4-16 depicts the operation of a spindle direct tool change. Figures 4-17 and 4-18 illustrate vertical machining centers employing spindle direct tool changers. When a tool change is initiated, the spindle is directed to the tool change location. The tool carousel indexes to the required tool slot, moves out of its "parked" position to the tooling position, and engages the toolholder that is in the spindle. The drawbar is then removed from the toolholder, and the tool carousel moves downward, removing the tool. The carousel then indexes to align the required tool with the spindle, and moves upward, inserting the tool into the spindle where the tool is secured. Finally, the carousel moves sideways away from the spindle, thus disengaging itself from the tool holder, and returns to its "parked" position. The tool change is now complete.

On some large vertical spindle machinery the procedure varies from the one just described. Tool carousels on very large machinery are too large to manipulate easily. Rather than move the carousel, the spindle is moved to the carousel and lowered over it to remove and insert the tools.

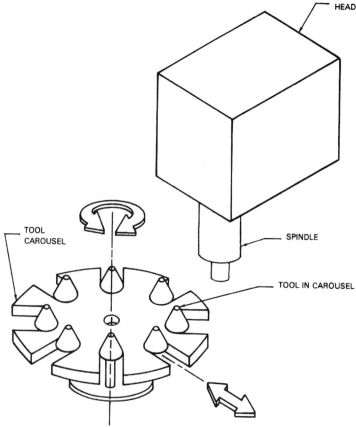

FIGURE 4-16
Spindle direct tool change

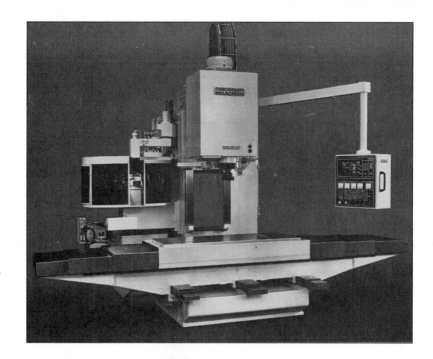

FIGURE 4-17
A vertical spindle machining center. Note the tool changer and carousel tool storage magazine. *(Photo courtesy of Cincinnati Milacron)*

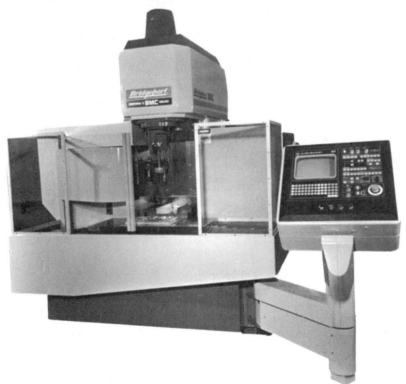

FIGURE 4-18
A vertical spindle machining center using carousel tool storage *(Photo courtesy of Bridgeport Machines Division of Textron Inc.)*

TOOL STORAGE

As with tool changers, there are as many tool storage systems as there are manufacturers. However, tool storage systems may be loosely grouped into two types: *carousel* and *matrix.*

Carousel Magazine

A carousel magazine stores the tools in a circular fashion. The machining centers pictured in Figures 4-3, 4-17, and 4-18 employ a tool carousel for tool storage. When a particular tool is called up, the carousel indexes to position the correct tool in the proper location for the tool changer to grab it. In addition to their use for spindle direct tool change on vertical spindle machining centers, carousels may be mounted on carts and moved to the proper spot as needed, such as when spindle direct tool change is employed on large equipment. They may also be mounted on the sides or backs of machines, depending on the type of tool changer used.

Matrix Magazine

Figures 4-2 and 4-14 picture machining centers employing matrix tool magazines. Figure 4-2 shows a single matrix magazine; Figure 4-14 shows a double. In either case, tool holder sockets are incorporated into long chains. When a tool is needed, the chain of sockets moves to position the correct tool socket in line with the tool changer. The advantage of the matrix magazine is that it is not limited to a circular configuration. In an oval configuration, for example, a matrix magazine can store a large number of tools in a limited amount of space.

TOOL LENGTH AND TOOL LENGTH OFFSET

Tools used for machining vary in length. When using three-axis NC machinery, some means to compensate for the differing tool lengths must be employed. There are two basic methods used to account for tool lengths: premeasuring the tools, and using the CNC controller's tool length compensation feature.

Preset Tool Method

One method of dealing with this problem is to set the tool to a specific length. This known length can then be added into the program's Z-axis coordinates to account for the tool. The setting of tools to a specific length is called *presetting*; the tools are called *preset tools*. Typically, the tool lengths are specified in an instruction sheet (developed by the programmer) that is sent to the shop floor for use in setting up the machine for a particular run. Sometimes a tool setup drawing is used. Special tool setting equipment is needed to measure the tools accurately. The cost of this equipment, and the labor necessary to set the tools, must be included in the cost of any numerical control system utilizing preset tooling. Preset tools make the replacement of broken or dull tools complicated, as such tools must be set to a specific length to function properly.

Tool Length Offset

The advent of CNC machinery has revolutionized tool setting by introducing the *programmable tool register.* A tool register is a memory spot in the computer where the length of a tool may be stored. When a particular tool is called up, the computer checks the tool register to see how much *offset* has been programmed for that tool (see the discussion on tool offset that follows). The MCU then shifts the Z-axis by the amount stored in the offset register. These offset figures usually are entered by the operator at the time

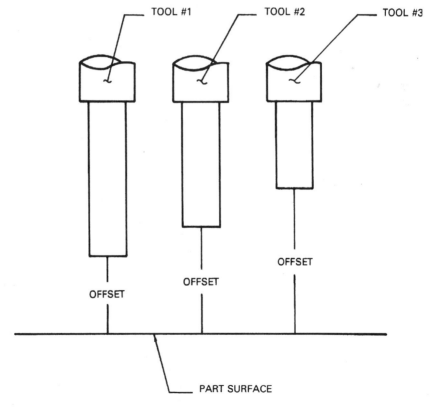

FIGURE 4-19
Tool length offset, difference of gage tool trim method

the machine is set up for the program run. There are three ways a machines tool length compensation can adjust for a tool length, called *trimming*.

1. Difference of gage tool trim.
2. Plus direction trim.
3. Minus direction trim.

Difference of Gage Tool Trim

This method is a variation of the preset tool method. The longest tool in the program is set to a predetermined length. This tool then becomes the gage length tool for the rest of the tools in the setup. The tool length offset for each tool will be the difference between the length of the tool and the

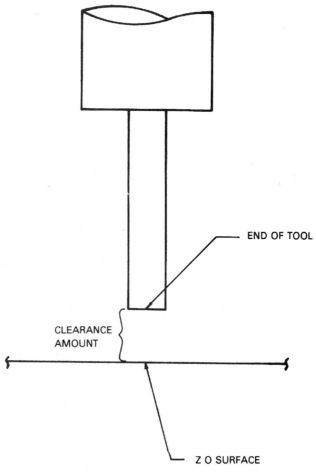

FIGURE 4-20
Tool clearance

gage tool. After a Z-zero point has been set, this longest tool is installed in the machine spindle. The table or machine head is then positioned with a specific distance between the tool and the workpiece. This distance is determined by either the programmer or setup man and must be sufficient to clear any clamps or other projections when the spindle is retracted (see Figure 4-20). To determine a particular tool offset, the tool is installed in the spindle, and the spindle lowered until the tool is at the desired Z-zero point on the part. The amount of offset for that tool will be displayed in the axis readout on the MCU. This offset amount, the distance from the tool to the part, is then entered in the MCU. The spindle can then be raised back to Z-Zero, the tool removed, and the procedure repeated for the next tool.

Each time a time a tool is called up by the program, the offset value for that tool is used to shift the original Z-zero point to the position on the part that the programmer desires as the Z-zero point for that tool. To fully retract the spindle, the tool offset is cancelled, shifting the Z-zero point back to its original position.

Plus Direction Trimming

This method does not rely on any tool being set to a specific length. The NC program is written without a specific tool length preprogrammed into the Z-axis coordinates. The length of each tool is simply measured, either on or out of the machine, and the entire length of the tool assembly is entered into

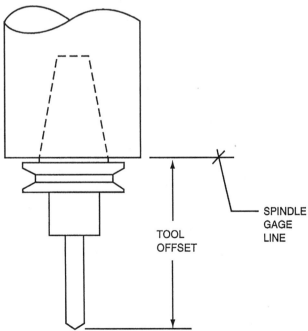

FIGURE 4-21
Tool length offset, plus direction trimming

the offset register for that tool. When the program calls up the tool, a command is issued in the program to compensate for the length of the tool in a Z-plus direction, toward the machine head (see Figure 4-21). A register number also is included to instruct the offset amount to the MCU. The MCU then shifts the programmed Z-axis coordinates by the amount contained in the offset register.

Minus Direction Trimming

This method is similar to the gage tool trim method in that the offset is the distance from the end of the tool to the desired Z-zero point. The difference is that the distance is not measured with respect to a gage tool's length. Rather, the distance is measured from the end of the tool with the spindle fully retracted to the Z-axis home position, to the desired Z-zero point on the part.

When the program calls up a tool, a command is issued, along with an offset register number, to trim the tool in the Z-minus direction, toward the machine table (see Figure 4-22). The MCU then shifts the Z-axis program coordinates toward the table by the amount in the offset register.

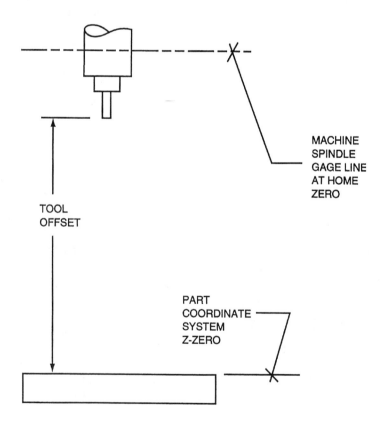

MACHINE
SPINDLE
GAGE LINE
AT HOME
ZERO

TOOL
OFFSET

PART
COORDINATE
SYSTEM
Z-ZERO

FIGURE 4-22
Tool length offset, minus direction trimming

SUMMARY

The important concepts presented in this chapter are:

- The speed, repeatability, and accuracy of a tool change greatly influence the efficiency of numerical control.
- There are two types of tool changes: manual and automatic.
- Machinery utilizing manual tool change generally incorporates some type of quick-change tooling system to facilitate the speed and accuracy of tool changes.
- Automatic tool changers are grouped into five categories: turret head, 180 degree rotation, pivot insertion, multi-axis, and spindle direct.
- Tool storage magazines are grouped into two types: carousel or matrix.
- Tool registers are places in the computer's memory to program tool offsets.

VOCABULARY INTRODUCED IN THIS CHAPTER

180-degree rotation tool changer
Automatic tool change (ATC)
Carousel tool magazine
Manual tool change
Matrix tool magazine
Multi-axis tool changer
Pivot insertion tool changer
Quick change tooling
Spindle direct tool changer
Tool length offset
Tool offset register
Turret head

REVIEW QUESTIONS

1. Why is tool changing so important in numerical control?
2. What are the two types of tool changes?
3. On what type of machinery are R-8 spindle tapers found?
4. On what type of machinery is an American Standard Machine Taper used?
5. What type of device is used on manual tool change machines to increase the speed of the tool change?
6. What are the five basic types of tool changers?

7. How does a 180-degree rotation tool changer work? How does a pivot insertion tool changer work?

8. What type of machinery is a spindle direct tool-changing system best suited for?

9. What are the two types of tool storage magazines?

10. What is a tool register?

11. What is a tool length offset?

12. Why are tool registers an improvement over other types of tool length solutions?

13. How does a programmer allow for tool length offsets in a part program?

14. What procedure is used by the operator to determine the tool length offsets?

CHAPTER 5

Programming Coordinates

OBJECTIVES Upon completion of this chapter, you will be able to:

- Explain what a hole operation is.
- Program hole operation coordinates using absolute and incremental positioning.
- Program milling coordinates using absolute and incremental positioning.

HOLE OPERATIONS

To understand how to program coordinates for hole operations, such as drilling, reaming, boring, and tapping, assume that the holes shown on the part drawing in Figure 5-1 are to be drilled using an absolute positioning machine. For hole #1, the coordinates are X0.7500, Y1.7500; for hole #2, the coordinates are X2.0000, Y0.2500; for hole #3, the coordinates are X3.0000, Y1.0000. Note that no plus or minus signs are given with any of these coordinates. If a coordinate is positive, no sign need be given; the machine will

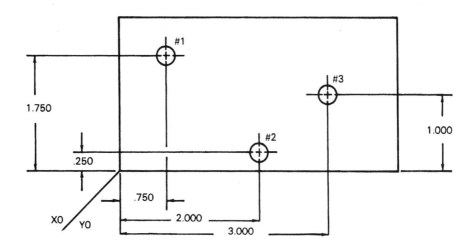

FIGURE 5-1
Datum dimensioned part drawing

101

assume a positive coordinate unless otherwise indicated. Looking at Figure 5-2, try to arrive at the coordinates to drill this part on an absolute positioning machine. The proper coordinates are as follows:

#1   X1.0000, Y0.5000   #5   X-1.0000, Y-0.5000
#2   X0.5000, Y1.0000   #6   X-0.5000, Y-1.0000
#3   X-0.5000, Yl .0000   #7   X0.5000, Y-1.0000
#4   X-1.0000, Y0.5000   #8   X1.0000, Y-0.5000

The same principles apply to the parts in Figures 5-1 and 5-2. The difference is that X0/Y0 is located at the center of the part in Figure 5-2. Notice that the signs of X and Y change as the coordinate locations move from quadrant to quadrant.

Figure 5-3 shows the same part as that in Figure 5-1 but delta dimensioned rather than datum dimensioned. Try to derive the proper coordinates to drill the holes in Figure 5-3, using an incremental positioning machine. The coordinates for the holes are as follows:

#1   X0.7500, Y1.7500
#2   X1.2500, Y-1.5000
#3   X1.0000, Y0.7500

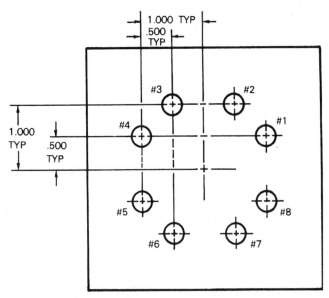

NOTES: 1) PART X0/Y0 IS CENTER OF PART
            2) FOR INCREMENTAL MOVES, THE SPINDLE IS ASSUMED
               CENTERED OVER X0/Y0 AT THE START OF PROGRAM
               SEQUENCE.

FIGURE 5-2

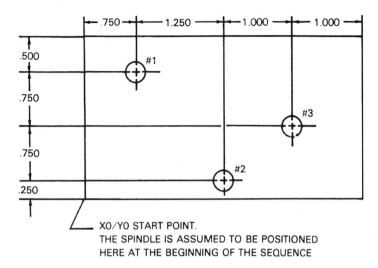

FIGURE 5-3
Delta dimensioned part drawing

Notice that the sign of Y was negative when moving to hole #2. Since incremental positioning was being used, hole #1 became the X0/Y0 point for the movement to hole #2. With incremental drawings, it is necessary to add and subtract dimensions in order to correctly program the part, even when using delta dimensioned drawings.

Referring again to Figure 5-2, assume that an incremental positioning machine is to be used. Determine the coordinates necessary to drill the part. The correct coordinates are:

#1	X1.0000, Y0.5000	#5	X0.0000, Y-1 .0000
#2	X-0.5000, Y0.5000	#6	X0.5000, Y-0.5000
#3	X-1.0000, Y0.0000	#7	X1.0000, Y0.0000
#4	X-0.5000, Y-0.5000	#8	X0.5000, Y0.5000

Even though this is a datum dimensioned drawing, it is often possible to program incrementally from it.

MILLING OPERATIONS

The system of coordinates presented thus far is used for centering a spindle over a particular location specified on a drawing. This means that when a coordinate location is given to the machine, the center of the spindle is sent to that location. In the case of milling cutters, this technique would cause a problem in that more than the correct amount of stock would be removed from the part (an amount equal to the radius of the cutter). When positioning the spindle for a milling operation, an allowance must be made for the radius of the cutter.

A .500-inch-diameter end mill is to be used to mill the part in Figure 5-4, and an absolute positioning mill will be used. Sending the cutter to X0/Y0 to begin a milling pass from location #1 to location #2 will remove an additional .250 inch of metal from the part that is called out in the drawing. To allow for the radius of the cutter, calculate the cutter coordinate by subtracting half the diameter of the cutter from the coordinate location in each axis. For location #1 the coordinates are X–0.2500, Y–0.2500. The coordinates for all four locations are as follows:

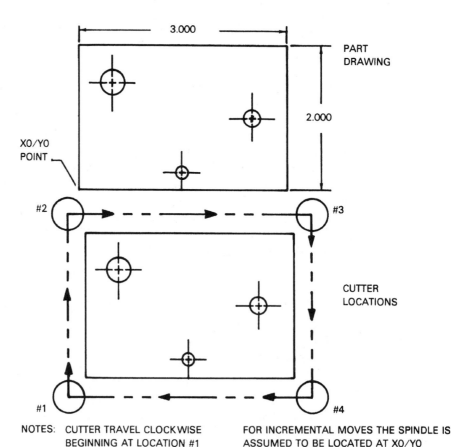

NOTES: CUTTER TRAVEL CLOCKWISE   FOR INCREMENTAL MOVES THE SPINDLE IS
         BEGINNING AT LOCATION #1   ASSUMED TO BE LOCATED AT X0/Y0
                                                    WHEN THE SEQUENCE STARTS

FIGURE 5-4

#1 X-0.2500, Y-0.2500
#2 X-0.2500, Y2.2500
#3 X3.2500, Y2.2500
#4 X3.2500, Y-0.2500

Assume that an absolute positioning machine is to be used to mill points indicated on the part drawing in Figure 5-5. The coordinates for this part are:

#1 X0.7500, Y0.7500
#2 X0.7500, Y1.2500
#3 X2.2500, Y1.2500
#4 X2.2500, Y0.7500

Try to determine the coordinates required to mill the parts in Figures 5-4 and 5-5, using an incremental positioning machine. The correct coordinates for Figure 5-4 are as follows:

#1 X-0.2500, Y-0.2500
#2 X0.0000, Y2.5000
#3 X3.5000, Y0.0000
#4 X0.0000, Y-2.5000

The correct coordinates for Figure 5-5 are as follows:

#1 X.07500, Y0.7500
#2 X0.0000, Y0.5000
#3 X1.5000, Y0.0000
#4 X0.0000, Y-0.5000

Movement of the Z axis is easier than that of the X or Y axis. To drill any of the parts examined in this chapter, all that is required is to give the Z axis a coordinate that would place the end of the tool through the part.

Assume that the zero point for the Z axis is the top of a .250-inch-thick part. A $^1/_4$-inch-diameter hole is to be used to drill a hole in the part. The coordinate for the Z axis would be the thickness of the part plus the length of the drill point. For a $^1/_4$-inch drill the coordinate would be Z−0.3250. The length of a drill point is calculated by multiplying the diameter of the drill by .3. In this case, the drill point is .075 inch long. This length added to the part depth (.250 inch) results in the .3250 length. In practice, it is wise to allow a small amount of additional movement to compensate for differences in drill point and part thickness tolerances. A movement toward the machine table would be a − Z movement. Movement toward the head of the machine would be a + Z movement. Chapter 7 covers three-axis milling and use of the Z axis in more detail. At this point, an understanding of the X and Y movements necessary to program coordinates will suffice.

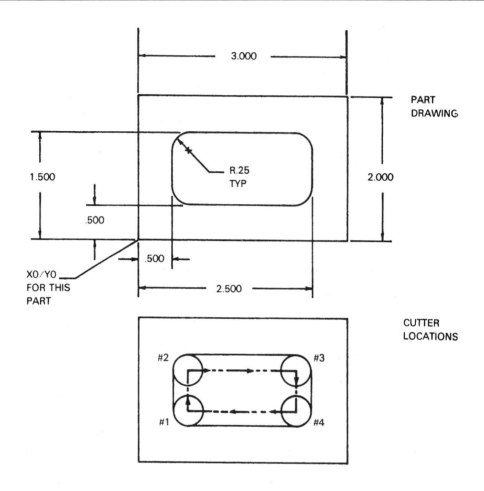

PART
DRAWING

CUTTER
LOCATIONS

NOTES: CUTTER TRAVEL - CLOCKWISE
BEGINNING AT LOCATION #1

FOR INCREMENTAL MOVES, THE SPINDLE IS
ASSUMED TO BE LOCATED AT X0/Y0
WHEN THE SEQUENCE STARTS

FIGURE 5-5

MIXING ABSOLUTE AND INCREMENTAL POSITIONING

CNC machines are capable of both incremental and absolute positioning. This gives the programmer a great deal of flexibility in programming parts. Assume that the part in Figure 5-6 is to be drilled using both absolute and incremental positioning. Hole #1 is to be drilled first, using absolute positioning; holes #2, #3, and #4 are to be drilled next, using incremental positioning; hole #5 is to be programmed next, using absolute positioning; and holes #6, #7, and #8 will be drilled using incremental positioning. Notice that the method of programming these coordinates is similar to the dimensioning used on the part print. Determine the coordinates to program the hole locations before looking at the following correct coordinates.

#1	X0.5000, Y-0.5000		#5	X2.7500, Y-2.0000
#2	X0.0000, Y-0.7500		#6	X0.0000, Y-0.7500
#3	X1.0000, Y0.0000		#7	X0.7500, Y0.0000
#4	X0.0000, Y0.7500		#8	X0.0000, Y0.7500

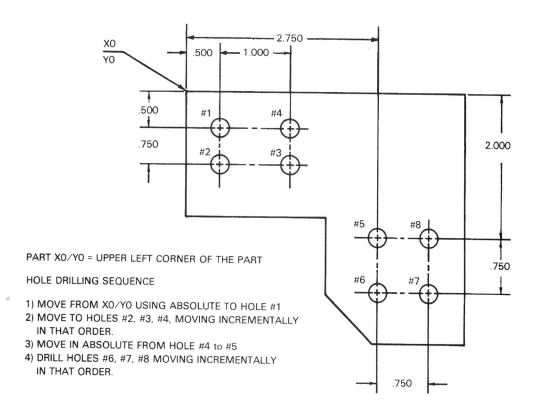

PART X0/Y0 = UPPER LEFT CORNER OF THE PART

HOLE DRILLING SEQUENCE

1) MOVE FROM X0/Y0 USING ABSOLUTE TO HOLE #1
2) MOVE TO HOLES #2, #3, #4, MOVING INCREMENTALLY IN THAT ORDER.
3) MOVE IN ABSOLUTE FROM HOLE #4 to #5
4) DRILL HOLES #6, #7, #8 MOVING INCREMENTALLY IN THAT ORDER.

FIGURE 5-6

METRIC COORDINATES

Some industries have converted all or part of their operations to metric units of measure. Most countries outside of the United States use metric measurement. It is advantageous, therefore, for companies with worldwide markets to use this system in manufacturing their products. Automobile manufacturers are but one example of a number of industries now converting to the metric system. It appears that both the inch and metric systems will be used for quite some time in the United States. Many experts agree that the United States will never fully convert to the metric system, but the numerical control programmer will have to deal with metric measures and should become familiar with their use in the shop.

The metric system in use today is called the *Système International d'Unites,* or the *SI* metric system. There are seven base units used in the metric system. Length is based on the *meter* (m), mass on the *kilogram* (kg), time on the *second (s),* electric current on the *ampere* (A), temperature on the *kelvin (K),* amount of substance on the *mole* (mol), and luminous intensity on the *candela* (cd). All metric units are built on a base-ten system. In the machine shop, measurement is based on the meter, which can be broken down into smaller units. A decimeter is 0.1 meter; a centimeter is 0.01 meter (0.1 decimeter); a millimeter is 0.001 meter (0.1 centimeter).

In the inch system, length measurement is based on the yard. Units smaller than a yard are built on fractions of a yard, foot, or inch, whichever is most compatible (one-half of a yard = $1^{1}/_{2}$ feet or 18 inches). In the machine shop, however, measurement is referenced to thousandths of inches. In the shop, 1" would be one inch; .500" however, is not thought of as five-tenths of an inch but, rather, as five-hundred thousandths of an inch. Therefore, .0005" is not usually called five ten-thousands of an inch, but five tenths, meaning five tenths of one-thousandth of an inch. When dealing with metric measurement in the shop, measurement is referenced to millimeters. One centimeter (1 cm) is not spoken of as one centimeter but is called ten millimeters (10 mm); one millimeter is approximately .0394 inch. Units smaller than one millimeter are also referenced in terms of millimeters; 0.01 is one-hundredth of a millimeter; 0.001 is one-thousandth of a millimeter; 0.001 inch is approximately .0254 millimeter; and .0001 inch is approximately 0.00254 millimeter. Many times a metric print tolerance will call for a two-place decimal to be held to + or − 0.02 mm. This roughly corresponds to holding an inch dimension to ± .001 inch.

Metric units are easy to work with as long as a company's commitment to metric conversion is carried all the way through from drafting room to tool crib. If metric cutters are available, working with metric dimensions is no problem. Modern CNC machinery has the capability to accept either metric or inch dimensions. The only difference in writing a program in metric versus inch measurements is that the coordinates are expressed differently. If inch tooling is used, it is necessary to convert the cutter sizes to metric units, so that proper milling coordinates can be programmed. To convert an inch dimension to a metric one, multiply the inch dimension by 25.4. To convert a metric dimension to one of inches, multiply the metric dimension by .03937 (or divide the metric dimension by 25.4).

Having learned the use of absolute and incremental positioning, and understanding how the Cartesian coordinate system works, a numerical control program may now be written.

SUMMARY

The important concepts presented in this chapter are:

- To program a hole location coordinate, the center line for the hole is used.
- To program a coordinate for milling operations, the coordinate for the location must include an appropriate allowance for the radius of the cutter.
- For absolute positioning, the datum reference plane remains the X0, Y0 point for all programmed moves.
- For incremental positioning, the current coordinate location is the X0, Y0 point for the next move.
- CNC machines are capable of mixing absolute and incremental positioning. This allows for flexibility in programming.
- Metric measurement in the machine shop is based on the millimeter, where .02 mm is roughly equivalent to .001 inch.
- To convert an inch dimension to millimeters, multiply the inch dimension by 25.4. To convert a metric dimension to inches, multiply the metric dimension by .03937, or divide the metric dimension by 25.4.

REVIEW QUESTIONS

1. What is a hole operation?
2. Where does the spindle centerline have to be programmed for a hole operation? For a milling operation?

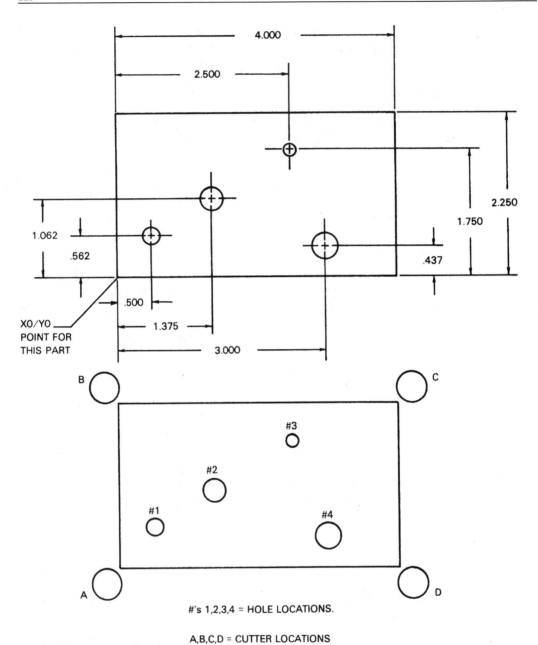

#'s 1,2,3,4 = HOLE LOCATIONS.

A,B,C,D = CUTTER LOCATIONS

FIGURE 5-7
Part drawing for review questions 3, 4, 5, and 6

(Questions #3, #4, #5, and #6 refer to Figure 5-7.)

3. What would the absolute coordinates for holes #2, #3, and #4 be?

4. What would the incremental coordinates be for holes #1, #3, and #4, moving to the holes in that order and starting at the lower left corner of the part?

5 Assume the spindle is positioned at hole #3. What would the incremental coordinates be to move from there to holes #2, #1, and #4, in that order? What would the absolute coordinates be?

6. Using a .625-inch-diameter end mill, what would be the four absolute coordinates necessary to mill the part periphery? What would be the incremental coordinates?

7. Assume that the hole patterns in Figure 5-8 are to be drilled using a CNC machine capable of both incremental and absolute positioning. Give the absolute coordinates to drill hole #1. Give the incremental coordinates to then drill holes a, b, c, and d, respectively. Give the absolute coordinate to drill hole #2, and the incremental coordinates to then drill holes e, f, g, and h, respectively.

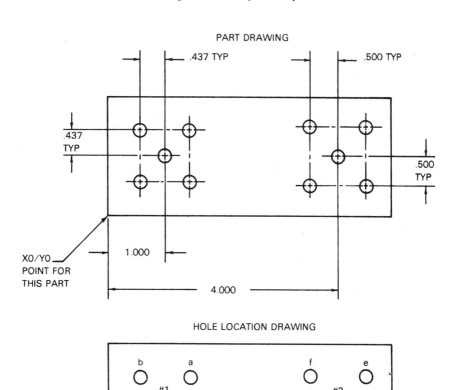

FIGURE 5-8
Part drawing for review question #7

8. Convert the following inch measurements to metric measurements.
   a. .500
   b. .4375
   c. .3125
   d. .125
9. Convert the following metric measurements to inch measurements.
   a. 0.02
   b. 0.005
   c. 2.5
   d. 8.0

CHAPTER 6

Two-Axis Programming

OBJECTIVES Upon completion of this chapter, you will be able to:

- Identify the basic parts of a CNC program.
- Describe the word address code format.
- Write simple two-axis programs in word address format to perform hole operations.
- Write simple two-axis milling programs using the word address format.
- Write simple two-axis programs that combine milling and hole operations.

INTRODUCTION

This text is concerned primarily with manual programming of CNC machinery. Each successive chapter will introduce a more advanced level of numerical control programming. For purposes of continuity, the same machine will be used for the next several chapters. The following point cannot be overemphasized: No two CNC machines program exactly alike. There are, however, similarities between them. By learning to write programs for this machine, only minimal effort will be required to program other CNC machines.

Programming in this text is done in a format called *word address*. It is the most common machine code format in use today. The machine programmed in this chapter is a vertical machining center equipped with a FANUC controller; it is a continuous path type machine. The program codes used on FANUC controls are similar to those used on other CNC controls (such as General Numeric and General Electric controls). This chapter deals specifically with two-axis programming. Chapter 7 will introduce programming utilizing the third machine axis. Although two-axis mill programming does not have much real world application, in an educational setting, learning two-axis programming first makes three-axis programming easier to understand.

PARTS OF A CNC PROGRAM

Regardless of the control or machine being programmed, all CNC programs consist of the same basic parts (or sections): (1) program startup section, (2) tool sequence safety (or startup) line, (3) tool load section, (4) tool motion section, (4) tool cancel section, and (6) the end of program section. The two-axis programs presented in this chapter will have some of these sections.

Program Startup

The program startup section serves to issue any commands required at the start of the tape only. For instance, setting the program to inch mode would only be required at the beginning of the program.

Tool Safety Block

The tool sequence safety block(s) serves to issue commands to cancel for any machine modes that could have been left active if the machine operator interrupted the tool cycle. By issuing a safety block the programmer and operator know the state of the machine at the beginning of the tool cycle.

Tool Load Blocks

The tool load section are those blocks of a tool sequence where the tool is placed in the spindle, either manually or by the machine's automatic tool changing mechanism, and the tool length compensation is turned on.

Tool Motion Blocks

The tool motion section contains the code for the actual cutting tool motion. It is where all the machining work is actually done.

Tool Cancel Blocks

The tool cancel section turns off the tool length compensation and returns the tool to the tool change position. All active cycle commands should be turned off in this section and the control left in a state ready to load the next tool.

End of Tape Blocks

The end of program blocks issue any commands necessary after all tool motion is complete, but before the program terminates. Often this section consists simply of the end of program code.

WORD ADDRESS FORMAT

As mentioned previously, the machine programmed in this chapter uses the word address format. Word address was developed as a tape programming format. Another name for word address is variable block format, so named because the program lines (blocks) may vary in length according to the information contained in them. Earlier tape formats required an entry for all possible machine registers. In these earlier formats, a zero was programmed as a null input if the register values were to be unaffected, but in work address, the blocks need only contain necessary information. Although developed as a tape format, word address is used as the format for manual data input on many CNC machines.

Addresses

The block format for word address is as follows:

N...G..X....Y...Z....I....J....K....F...H..S....T..M..

Only the information needed on a line need be given. Each of the letters is called an address (or word). The various words are as follows:

N – The block sequence number. An N number is used to number the lines of NC code for operator and/or programmer reference. N numbers are ignored by the controller during program execution. Most NC controls allow a block to be searched for by the sequence number for editing or viewing purposes.

G – Initiates a preparatory function. Preparatory functions change the control mode of the machine. Examples of preparatory functions are rapid/feedrate mode, drill mode, tapping mode, boring mode, and circular interpolation. Preparatory functions are called prep functions, or more commonly G codes.

X – Designates an X-axis coordinate. X also is used to enter a time interval for a timed dwell on FANUC and FANUC-style controls.

Y – Designates a Y-axis coordinate.

Z – Designates a Z-axis coordinate.

I – Identifies the X-axis arc vector (the X-axis center point of an arc).

J – Identifies the Y-axis arc vector (the Y-axis center point of an arc).

K – Identifies the Z-axis arc vector (the Z-axis center point of an arc).

S – Sets the spindle RPM.

H – Specifies the tool length compensation register.

F – Assigns a feedrate.

T – Specifies the standby tool (the tool to be used in the next tool change).

M – Initiates miscellaneous functions (M functions). M functions control auxiliary functions such as the turning on and off of the spindle and coolant, initiating tool changes, and signaling the end of a program.

Other words used in word address will be explained as they are used. A list of EIA codes for word address is contained in Appendix 1.

Leading And Trailing Zeros

Before presenting program explanations, a discussion of leading and trailing zeros is in order. In the shop and on part drawings, dimensions often contain trailing zeros. For example, the dimension .500 contains two trailing zeros. On part dimensions, the trailing zeros are necessary to communicate the significance of a particular dimension (that is, three-place vs. two-place decimals). Sometimes, for the sake of clarity, a leading zero is used, as in 0.500. Early NC equipment required the use of leading and/or trailing zeros in specifying any coordinate. Many CNC controllers do not require the use of either leading or trailing zeros; thus, .500 may be entered as .5 on these machines. Similarly 1.000 may be entered as 1. (the decimal point only following the number). The CNC machine will locate all programmed coordinates within the resolution of the machine. The programs in this chapter reflect this practice of omitting leading and trailing zeros. Not all controllers allow this practice; the FANUC-style controllers used in this text do.

DRILLING EXAMPLE – ABSOLUTE POSITIONING

Figure 6-1 is a part to be programmed. Figure 6-2 is a metric version of the same part. Figure 6-3 contains a program, written in the word address format, to drill the part in Figure 6-1. Figure 6-4 contains the program to drill the part in Figure 6-2. Comments within the body of the program are placed inside parentheses. The open parenthesis "(" is called the control out character. The closed parenthesis ")" is called the control in character. All text between the control out and control in characters is ignored by the control.

The programs in 6-3 and 6-4 contain the following sections

1. Program and tools startup, sequence block N010.
2. Tool motion blocks, N020 through N070.
3. Tool cancellation block, N080.
4. End of program blocks, N090.

Program Explanation

%

The percent sign (%) is used on FANUC style controllers to signal the beginning of a program sequence to an input device. Its primary purpose is to signal the program read port on the machine when to begin reading program characters being input off of disk, DNC file transfer, or punched tape. It is also called a tape start code, since its use originated on punched tape.

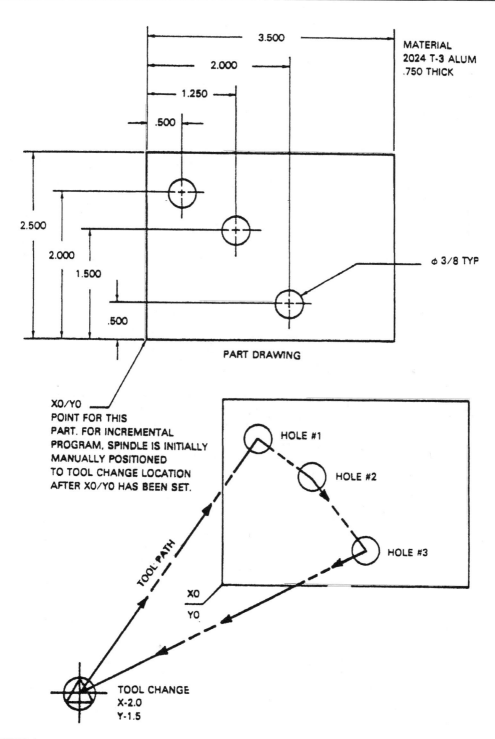

FIGURE 6-1
Hole operations part drawing, inch

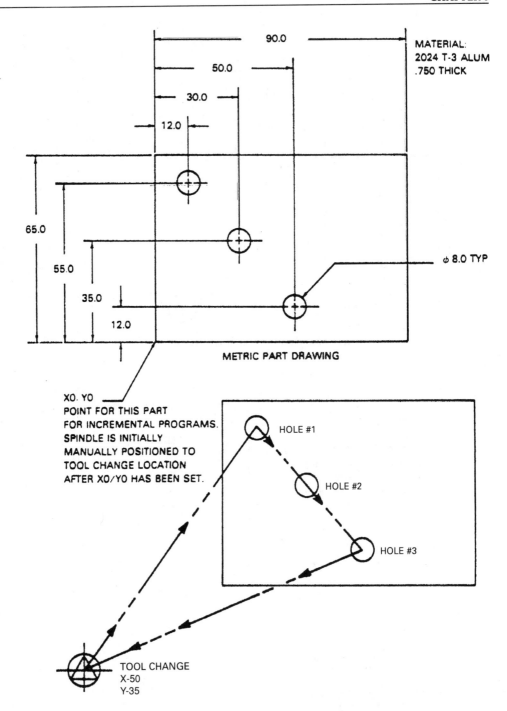

FIGURE 6-2
Hole operations part drawing, metric

```
%
O6003
(*******************************)
(* PROGRAM X0/Y0 = TOOL CHANGE)
(* SET TOOL CHANGE X-2.0 Y-1.5 PRIOR TO)
(* STARTING FIRST CYCLE.)
(* *******************************)
(TOOL 1 - 3/8 DRILL)
(SPINDLE SPEED 2500 RPM)
N010 G00 G70 G90 M06
N020 X.5 Y2.
N030 M00
N040 X1.25 Y1.5
N050 M00
N060 X2. Y.5
N070 M00
N080 X-2. Y-1.5
N090 M30
%
```

FIGURE 6-3

Drilling program, inch absolute positioning, for the part in Figure 6-1

```
%
O6004
(* *******************************)
(* PROGRAM X0/Y0 = TOOL CHANGE)
(* SET TOOL CHANGE X-50. Y-35. PRIOR TO)
(* STARTING FIRST CYCLE.)
(* *******************************)
(TOOL 1 - 3/8 DRILL)
(SPINDLE SPEED 2500 RPM)
N010 G00 G71 G90 M06
N020 X12. Y55.
N030 M00
N040 X30. Y35.
N050 M00
N060 X50. Y12.
N070 M00
N080 X-50. Y-35.
N090 M30
%
```

FIGURE 6-4

Drilling program metric absolute positioning. for the part in Figure 6-1

O6003

This is the program number. All FANUC-style controllers (and many other controls) use the letter "O" to designate the start of a program in controller memory. A colon ":" can be substituted for the letter "O."

N010

N010 – This is the sequence number. Each block (line) of the program begins with a unique number, prefaced with the N address. This program follows the practice of numbering the blocks by tens. This particular control does not care if the numbers are not in order. Numbering the program by tens allows any blocks inserted by way of in-shop editing to be numbered consecutively. Within this text, both numbering by tens and sequential block numbering are used.

G00 – Puts the machine in rapid traverse mode. All machine moves made while G00 is active will be made at maximum speed.

G70 – Puts the machine in inch input mode. All program coordinates will be read in decimal inches.

G90 – Puts the machine in absolute positioning mode.

M06 – Issues the tool change command. In this two-axis milling program, it causes the control to stop and wait for the operator to install a toolholder in the spindle.

N020

N020 – The block sequence number.

X/Y coordinates – Cause the machine to position the tool from the tool change position to hole #1.

N030

N030 – Block sequence number.

M00 – Program stop command. This halts the program execution, allowing the operator to drill the holes.

N040

N040 – Block sequence number.

X/Y coordinates – to move from hole #1 to hole #2.

N050

N050 – Block sequence number.

M00 – Program stop command. This halts the program so that hole #2 can be drilled.

N060

N060 – Block sequence number.

X/Y coordinates – To move from hole #2 to hole #3.

N070

N070 – Block sequence number.

M00 – Program stop command. Hole #3 is drilled.

N080

N080 – Block sequence number.

X/Y coordinates – to move from hole #3 to tool change.

N090

N090 – Block sequence number.

M30 – Signals that the program has ended and resets the computer's memory to the start of the program.

%

The percent sign (%) is used on FANUC-style controllers as an end of program marker. This second percent sign signals the program read port on the CNC control to stop reading characters from the input device. It is also called a tape stop code, since its use originated with punched tape.

DRILLING EXAMPLE – INCREMENTAL POSITIONING

Figure 6-5 is a program to drill the part in Figure 6-1 using incremental positiong. Figure 6-6 is a program to drill the metric part in Figure 6-2 using incremental positioning. The program sequence used in these examples is identical to that used in Figures 6-3 and 6-4. Only the coordintates differ, reflecting the change in positioning system.

```
%
O6005
(* ********************************)
(* PROGRAM X0/Y0 = TOOL CHANGE)
(* SET TOOL CHANGE X-2.0 Y-1.5 PRIOR TO)
(* STARTING FIRST CYCLE.)
(* ********************************)
(TOOL 1 - 3/8 DRILL)
(SPINDLE SPEED 2500 RPM)
N010 G00 G70 G91 M06
N020 X2.5 Y3.5
N030 M00
N040 X.75 Y-.5
N050 M00
N060 X.75 Y-1.
N070 M00
N080 X-4. Y-2.
N090 M30
%
```

FIGURE 6-5

Drilling program, inch incremental positioning, for the part in Figure 6-1

```
%
O6006
(* *************************************)
(* PROGRAM X0/Y0 = TOOL CHANGE)
(* SET TOOL CHANGE X-50. Y-35. PRIOR TO)
(* STARTING FIRST CYCLE.)
(* *************************************)
(TOOL 1 - 3/8 DRILL)
(SPINDLE SPEED 2500 RPM)
N010 G00 G71 G91 M06
N020 X62. Y90.
N030 M00
N040 X18. Y-20.
N050 M00
N060 X20. Y-23.
N070 M00
N080 X-100. Y-47.
N090 M30
%
```

FIGURE 6-6
Drilling program, metric incremental positioning for the part in Figure 6-2

%
   Percent sign (%) – program start code.

O6005
   This is the program number (the "O" number).

N010
   N010 – The block sequence number.
   G00 – Puts the machine in rapid traverse mode. All machine moves made while G00 is active will be made at maximum speed.
   G70 – Puts the machine in inch input mode. All program coordinates will be read in decimal inches.
   G91 – Puts the machine in incremental positioning mode. In the absolute positioning example, the G90 code was used.
   M06 – Issues the tool change command. In this two-axis milling program, it causes the control to stop and wait for the operator to install a toolholder in the spindle.

N020
   N020 – The block sequence number.
   X/Y coordinates – Causes the machine to position the tool from the tool change position to hole #1.

N030

N030 – Block sequence number.

M00 – Program stop command. This halts the program execution, allowing the operator to drill the holes.

N040

N040 – Block sequence number.

X/Y coordinates – to move from hole #1 to hole #2.

N050

N050 – Block sequence number.

M00 – Program stop command. This halts the program so that hole #2 can be drilled.

N060

N060 – Block sequence number.

X/Y coordinates – To move from hole #2 to hole #3.

N070

N070 – Block sequence number.

M00 – Program stop command. Hole #3 is drilled.

N080

N080 – Block sequence number.

X/Y coordinates – to move from hole #3 to tool change.

N090

N090 – Block sequence number.

M30 – Signals that the program has ended and resets the computer's memory to the start of the program.

%

Percent sign (%) – Program stop code.

MILLING EXAMPLE 1

Assume the part in Figure 6-7 is to be milled. The part is an aluminum casting which requires that only the length and width be machined. Figure 6-8 is a metric part. The part setup drawing is Figure 6-9. Clamping will be done through the center hole. Two passes around the part will be made, a roughing pass and a finishing pass. .010 of stock will be left for the finish pass (.25mm metric version).

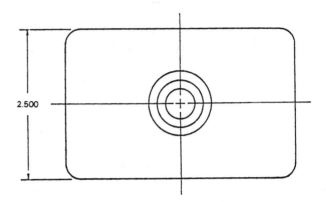

MATERIAL: ALUMINUM CASTING
NOTE: ONLY PERTINENT DIMENSIONS GIVEN

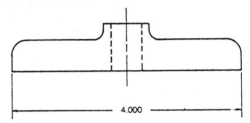

FIGURE 6-7
Milling part drawing, inch

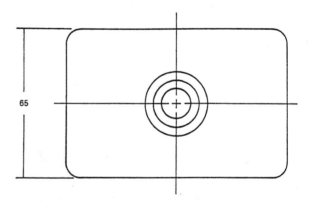

MATERIAL: ALUMINUM CASTING
NOTE: ONLY PERTINENT DIMENSIONS GIVEN

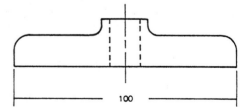

FIGURE 6-8
Milling part drawing, metric

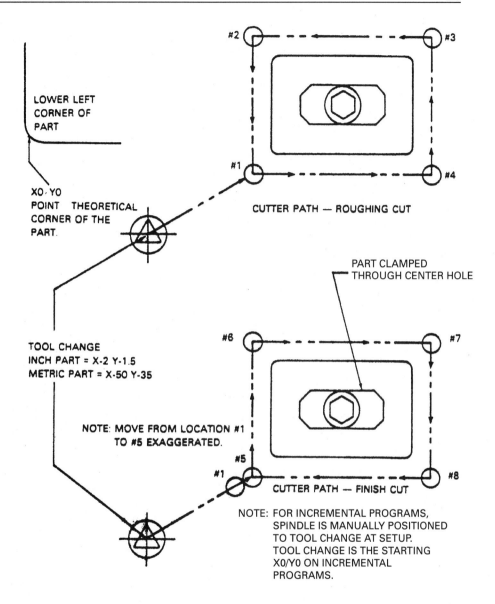

LOWER LEFT
CORNER OF
PART

X0, Y0
POINT   THEORETICAL
CORNER OF THE
PART.

CUTTER PATH — ROUGHING CUT

TOOL CHANGE
INCH PART = X-2 Y-1.5
METRIC PART = X-50 Y-35

PART CLAMPED
THROUGH CENTER HOLE

NOTE: MOVE FROM LOCATION #1
TO #5 EXAGGERATED.

CUTTER PATH — FINISH CUT

NOTE: FOR INCREMENTAL PROGRAMS,
SPINDLE IS MANUALLY POSITIONED
TO TOOL CHANGE AT SETUP.
TOOL CHANGE IS THE STARTING
X0/Y0 ON INCREMENTAL
PROGRAMS.

FIGURE 6-9
Setup drawing for the part in Figures 6-7 and 6-8

Two programs, one nonmetric and one metric, written using absolute positioning will be presented first; the programs are contained in Figures 6-11 and 6-12. A .500 inch diameter end mill will be used in the nonmetric program and a 5mm diameter metric end mill will be used in the metric version.

Up and Down Milling

When milling cuts are programmed, it is important to understand the difference between up and down milling. Figure 6-10 illustrates these two machining practices. Notice that in up milling (also called *conventional milling*), the cutter forces acting on the part try to lift the part up off the table, hence the name *up milling*. In down milling the force of the cutter tries to push the part downward onto the table, thus the name *down milling*. Down milling also is referred to as climb cutting, because the cutter is trying to climb up on top of the part.

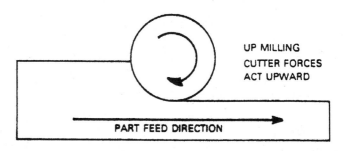

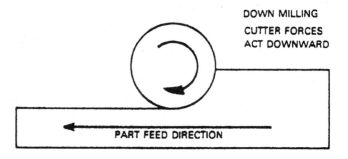

FIGURE 6-10
Up milling and down milling

Up milling is used for cutting most ferrous materials, brass and bronze, and for roughing cuts on aluminum and aluminum alloys. Down milling is used for finishing cuts on aluminum and aluminum alloys. It also is occasionally used for finishing cuts on other metals, if conditions warrant it. Down milling requires less power for a particular cut, but places more stress on the machine slides and ball screws than does up milling. Exactly when to use up and down milling is something that must be learned through experience as it depends not only on the machine available, but also on the cutting tools, coolant, and workpiece materials.

MILLING EXAMPLE 2

The part pictured in Figure 6-7 will now be milled. Figure 6-11 is the program in absolute positioning. Figure 6-12 is the metric version to mill the part in Figure 6-8. Notice that only an X or Y coordinate, rather than an X/Y pair of coordinates, is used in some lines. If the machine is already positioned in one of its axes, a coordinate for that axis need not be given. No movement is to take place, therefore, the second coordinate is not required.

```
%
O6011
(*****************************************************)
(* THIS PROGRAM USES ABSOLUTE POSITIONING)
(* X/Y ORIGIN IS LOWER LEFT CORNER OF PART)
(* PLACE 1/2 END MILL IN SPINDLE PRIOR TO CYCLE START)
(*****************************************************)
(SET PARAMETERS TO RAPID - INCH INPUT -ABS. POS.)
N010G00G70G90
(AT PROG. STOP - LOWER SPINDLE AND CLAMP)
N020X-.26Y-.26
N030M00
(BEGIN ROUGH MILL CUT AT FEEDRATE)
N040G01X4.26F20.
N050Y2.76
N060X-.26
N070Y-.26
(BEGIN FINISH MILL CUT)
N080X-.25Y-.25
N090Y2.75
N100X4.25
N110Y-.25
N120X-.25
(AT PROG. STOP UNCLAMP AND RAISE SPINDLE)
N130M00
(RETURN TO TOOL CHANGE LOCATION AND END CYCLE)
N140G00X-2.Y-1.5
N150M30
%
```

FIGURE 6-11

Milling program, inch absolute positioning for the part in Figure 6-7

```
%
06012
(*************************************************)
(* THIS PROGRAM USES ABSOLUTE POSITIONING)
(* X/Y ORIGIN IS LOWER LEFT CORNER OF PART)
(* PLACE 1/2 END MILL IN SPINDLE PRIOR TO CYCLE START)
(*************************************************)
(SET PARAMETERS TO RAPID - INCH INPUT -ABS. POS.)
N010G00G71G90
(AT PROG. STOP - LOWER SPINDLE AND CLAMP)
N020X-2.75Y-2.75
N030M00
(BEGIN ROUGH MILL CUT AT FEEDRATE)
N040G01X102.75F500.
N050Y67.75
N060X-2.75
N070Y-2.75
(BEGIN FINISH MILL CUT)
N080X-2.5y-2.5
N090Y67.5
N100X102.5
N110Y-2.5
N120X-2.5
(AT PROG. STOP UNCLAMP AND RAISE SPINDLE)
N130M00
(RETURN TO TOOL CHANGE LOCATION AND END CYCLE)
N140G00X-50.Y-35.
N150M30
%
```

FIGURE 6-12

Milling program, metric absolute positioning for the part in Figure 6-8

%

Percent sign (%) – Program start code.

06011

The program number.

N010

N010 – The block sequence number.

G00 – Puts the machine in rapid traverse mode.

G70/G71 – Puts the machine in inch or metric mode.

G90 – Selects absolute positioning.

N020

N020 – Block sequence number.

X/Y coordinates – To move to location #1.

N030

N030 – Sequence number.

M00 – Program stop. This command halts the program execution to allow the machine operator to lower and clamp the spindle.

N040

N040 – Sequence number.

G01 – Puts the machine in feedrate mode (also called linear interpolation mode as explained later in Chapter 9).

X coordinate – Moves spindle from position #1 to #4.

F20. – Specifies a feedrate of 20 inches per minute is to be used during feedrate moves.

N050

N050 – Sequence number.

Y coordinate – Moves spindle from position #4 to #3.

N060

N060 – Sequence number.

X coordinate – Moves spindle from position #3 to #2.

N070

N070 – Sequence number.

Y coordinate – Moves spindle from position #2 to #1.

N080

N080 – Sequence number.

X/Y coordinates – Move spindle from position #1 to #5.

N090

N090 – Sequence number.

Y coordinate – Moves spindle from position #5 to #6.

N100

N100 – Sequence number.

X coordinate – Moves spindle from position #6 to #7.

N110

N110 – Sequence number.

Y coordinate – Moves spindle from position #7 to #8.

N120

N120 – Sequence number.

X coordinate – Moves spindle from position #8 to #5.

N130

N130 – Sequence number.

M00 – Program stop code. This command halts the program execution, allowing the machine operator to unclamp and raise the spindle.

N140

N140 – Sequence number.

G00 – Places the machine in rapid traverse mode.

X/Y coordinates – Move the spindle from position #5 to the tool change location.

N150

N150 – Sequence number.

M30 – Signals end of program.

%

Percent sign – Program stop code.

MILLING AND DRILLING EXAMPLE

The part in Figure 6-13 is to be milled and drilled on a two-axis machine; it is an aluminum casting. The part will be clamped to the table through the as-cast slot. Figure 6-14 illustrates the cutter path used to mill the part. Figure 6-15 illustrates the spindle path for drilling the part. The part program is given in Figure 6-16.

This program illustrates a different method of using sequence numbers. Since sequence numbers are ignored by the control except for block searching purposes, the programmer may use them as an aid to reading a program. In this program, the beginning of each tool starts with a sequence number corresponding to the tool. N001 begins tool 1 sequence, N002 begins tool 2 sequence, and N003 begins tool 3 sequence. The sequence number is the only code in the beginning block. Every block used with tool 1 uses a 100 series sequence number, tool 2 uses 200 series numbers, and tool 3 uses 300 series numbers.

The machine used to machine this part is equipped with automatic spindle speed capability. Whereas in the previous examples the operator turned the spindle on and off, this machine will control the spindle speed and direction via CNC commands. Two codes are used in this example to accomplish this: S codes and M03. The S code sets the spindle RPM. The code M03 turns the spindle on in a clockwise direction.

Program Explanation

%

Percent sign – Program start code.

O6018

Program "O" number.

N001

N001 – Block sequence number, used as a separator line.

N100

N100 – Sequence number.

G00 – Selects rapid traverse mode.

G90 – Selects absolute positioning mode.

G70 – Selects inch input mode.

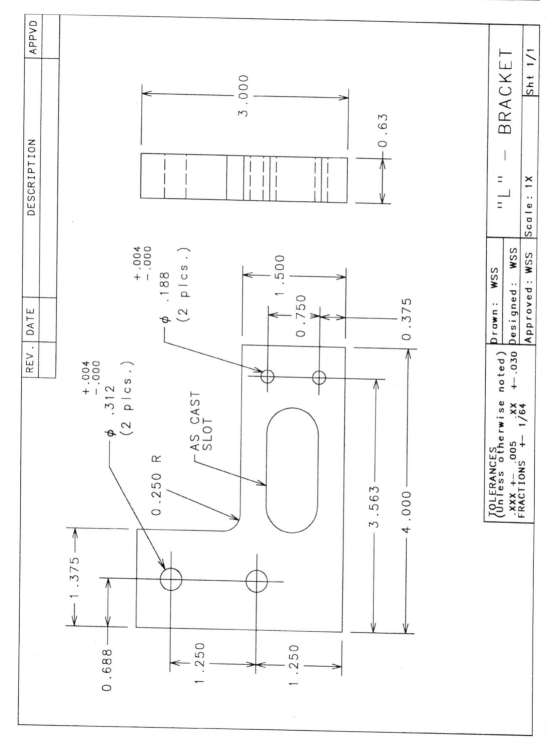

FIGURE 6-13
Milling and drilling part drawing

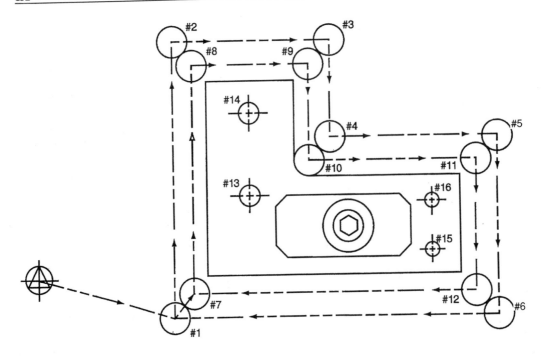

FIGURE 6-14
Cutter path for milling the part in Figure 6-13

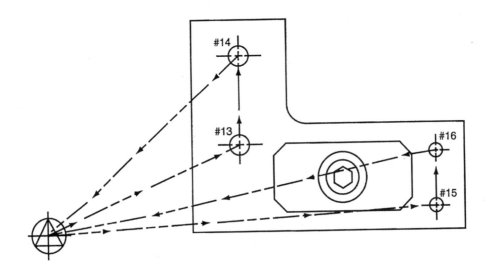

FIGURE 6-15
Cutter path for drilling the part in Figure 6-13

```
%
06016
(* *******************)
(* L-BRACKET  06/09/93)
(* *******************)
N001
(* ***********)
(* TOOL NO. 1)
(* .500 DIA. END MILL)
(* ROUGH/FINISH PART PERIPHERY)
(* ***********)
N100G00G90G70
N101X.26Y.26S1200M03
N102M00
(LOWER AND CLAMP SPINDLE)
(BEGIN ROUGH MILL PASS - LEAVE .01 STK/SIDE)
N103G01Y3.26F7.2
N104X1.635
N105Y1.76
N106X4.26
N107Y-.26
N108X-.26
(BEGIN FINISH MILL PASS)
N109X-.25Y-.25
N110Y3.25
N111X1.625
N112Y1.75
N113X4.25
N114Y-.25
N115X-.25
N116M00
(UNCLAMP AND RAISE SPINDLE)
(AT CYCLE START RETURNS TO TOOL CHANGE)
N117G00X-10.Y0.
N118M00

N002
(* ***********)
(* TOOL NO. 2)
(* 5/16 STUB DRILL)
(* DRILL THE .312 DIA. HOLES)
(* ***********)
N200G00G90G70
N201X.688Y1.25S1500M03
N202M00
(DRILL HOLE)
N203Y2.5
N204M00
(DRILL HOLE)
N205X-10.Y.0
N206M00

N003
(* ***********)
(* TOOL NO. 3)
(* 3/16 STUB DRILL
(* DRILL .188 DIA. HOLES)
(* ***********)
N300G00G90G70
N301G00X3.563Y.375S2000M03
N302M00
(DRILL HOLE)
N303Y1.125
N304M00
(DRILL HOLE)
N305G00X-10.Y.0
N306M30
%
```

FIGURE 6-16

Program to mill and drill the part in
Figure 6-13

N101

N101 – Sequence number.
X/Y coordinates – Positions spindle at location #1.
S1200 – Selects a spindle speed of 1200 to be used for milling the part.
M03 – Turns spindle on clockwise.

N102

N102 – Sequence number.
M00 – Program stop command. Halts program execution so operator can lower and clamp the spindle.

N103

N103 – Sequence number.
G01 – Selects feedrate mode.
Y coordinate – Moves spindle from position #1 to #2.
F7.2 – Sets the feedrate to 7.2 inches per minute.

N104

N104 – Sequence number.
X coordinate – Moves spindle from position #2 to #3.

N105

N105 – Sequence number.
Y coordinate – Moves spindle from position #3 to #4.

N106

N106 – Sequence number.
X coordinate – Moves spindle from position #4 to #5.

N107

N107 – Sequence number.
Y coordinate – Moves spindle from position #5 to #6.

N108

N108 – Sequence number.
X coordinate – Moves spindle from position #6 back to #1.

N109

N109 – Sequence number.
X/Y coordinates – Moves spindle from position #1 to #7.

N110

N110 – Sequence number.
Y coordinate – Moves spindle from position #7 to #8.

N111

N111 – Sequence number.
X coordinate – Moves spindle from position #8 to #9.

N112

N112 – Sequence number.
Y coordinate – Moves spindle from position #9 to #10.

N113

N113 – Sequence number.

X coordinate – Moves spindle from position #10 to #11.

N114

N114 – Sequence number.

y coordinate – Moves spindle from position #11 to #12.

N115

N115 – Sequence number.

X coordinate – Moves spindle from position #12 back to #7.

N116

N116 – Sequence number.

M00 program stop code – Halts program execution so operator can unclamp and raise the spindle.

N117

N117 – Sequence number.

X/Y coordinates – Move the spindle from position #7 back to the tool change location.

N118

N118 – Sequence number.

M00 – Halts program for a tool change.

N002

N002 – Block sequence number used as a separator line for tool 2.

N200

N200 – Sequence number.

G00 – Selects rapid traverse mode.

G90 – Selects absolute positioning mode.

G70 – Selects inch input mode.

N201

N201 – Sequence number.

X/Y coordinates – Position spindle at location #13.

S1500 – Selects a spindle speed of 1500 rpm to be used for drilling the holes.

N202

N202 – Sequence number.

M00 – Program stop command. Halts program execution so operator can drill the hole.

N203

N203 – Sequence number.

Y coordinate – Moves spindle from hole #13 to #14.

N204

N204 – Sequence number.

M00 – Program stop command. Halts program execution so operator can drill the hole.

N205

N205 – Sequence number.
X/Y coordinates – Move spindle back to the tool change position.

N206

N206 – Sequence number.
M00 – Halts program for a tool change.

N003

N003 – Block sequence number used as a separator line for tool 3.

N300

N300 – Sequence number.
G00 – Selects rapid traverse mode.
G90 – Selects absolute positioning mode.
G70 – Selects inch input mode.

N301

N301 – Sequence number.
X/Y coordinates – Position spindle at location #15.
S2000 – Selects a spindle speed of 2000 rpm to be used for drilling the holes.
M03 – Turns spindle on clockwise.

N302

N302 – Sequence number.
M00 – Program stop command. Halts program execution so operator can drill the hole.

N303

N303 – Sequence number.
Y coordinate – Moves spindle from hole #15 to #16.

N304

N304 – Sequence number.
M00 – Program stop command. Halts program execution so operator can drill the hole.

N305

N305 – Sequence number.
X/Y coordinates – Move spindle back to the tool change position.

N306

N306 – Sequence number.
M30 – Signals end of program.

%

Percent sign – Program stop code.

SUMMARY

The important concepts presented in this chapter are:
- An NC or CNC program consists of six basic parts

    1. Program startup section
    2. Tool sequence safety line
    3. Tool load (or tool change) section
    4. Tool motion sequence
    5. Tool cancel section
    6. End of program section.

- In word address format, each CNC command is called a word. Each word begins with an alpha address which identifies the command's function. The address is followed by a numeric value. Some values are used to set machine modes. Others are used to specify positioning coordinates.
- The spindle must be positioned safely out of the way at the end of the program, to allow safe loading and unloading of the workpiece. This is accomplished in both the milling and drilling examples by sending the spindle back to its tool change location at the end of the program.
- Incremental programs differ from absolute programs only in the coordinates used. Programs in absolute and incremental positioning use the same programming logic. In incremental positioning, it is imperative that the machine start and stop in the same location. Failure to program for this will result in incorrect positioning for the second cycle.
- To perform hole operations, it is necessary to position the spindle over the centerline of the hole.
- A program stop command is used at hole locations to halt the program and enable the operator to drill the hole.
- When programming coordinates for milling, an allowance must be made for the size of the cutter.

VOCABULARY INTRODUCED IN THIS CHAPTER

Addresses
End of tape blocks
Leading zero
Program startup blocks
Tool cancel blocks
Tool load blocks
Tool motion blocks
Tool safety blocks
Trailing zero
Two-axis programming

REVIEW QUESTIONS

1. What are the six basic parts of a CNC program?
2. What do the following addresses stand for in word address format: X, Y, G, M, S, F, N?
3. What is a preparatory function?
4. What are miscellaneous functions?

(Questions 5 and 6 refer to the part in Figure 6-17. The cutter path drawing is given in Figure 6-18.)

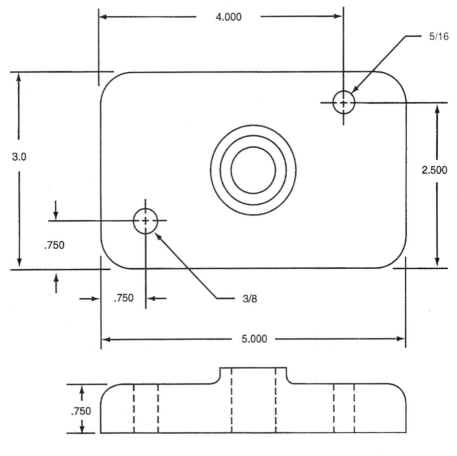

INSTRUCTIONS   1) MILL AND DRILL PART
                2) USE LOWER LEFT CORNER FOR XO YO

FIGURE 6-17
Part drawing for Review Questions 5 and 6

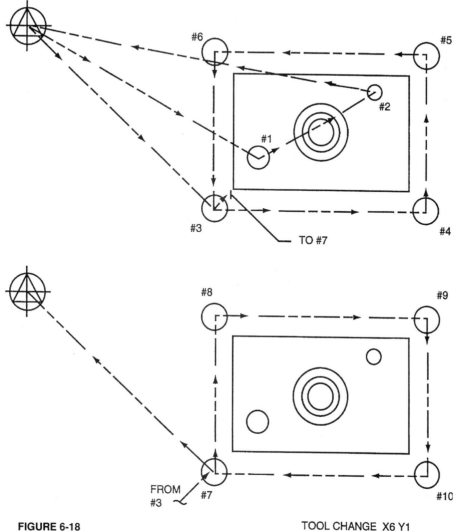

FIGURE 6-18
Cutter path for Figure 6-17

TOOL CHANGE X6 Y1
XO YO LOWER LEFT CORNER

5. Write a program in word address format to mill and drill the part using absolute positioning.
6. Write a program in word address format to mill and drill the part using incremental positioning.

(Question 7 refers to the part in Figure 6-19. The cutter path drawing is given in Figure 6-20.)

7. Write a program in word address format to mill and drill the part using absolute positioning.

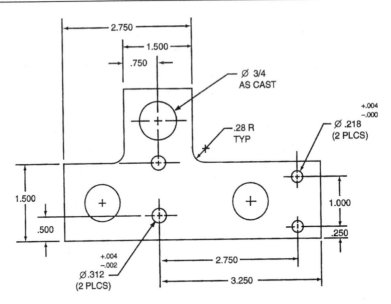

FIGURE 6-19
Part drawing for Review Question 7

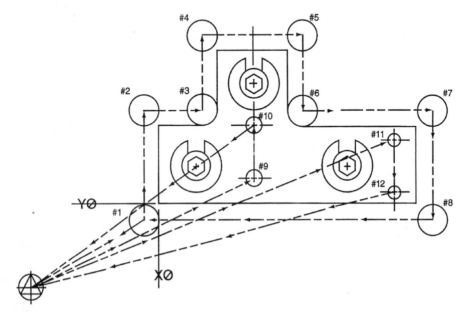

FIGURE 6-20
Cutter path for Figure 6-19

CHAPTER 7

Three-Axis Programming

OBJECTIVES Upon completion of this chapter, you will be able to:

- Write simple programs to perform hole operations using three machine axes.
- Explain what a canned cycle is.
- Explain the difference between initial level and reference level on CNC machinery.
- Explain the difference between a modal and nonmodal command.
- Write simple programs to perform milling operations using three machine axes.
- Write simple programs involving a machine indexer.

INTRODUCTION

In the previous chapter, only two axes, the X and Y axis, were used to machine the part. Programming using only two axes is called two-axis programming. In this chapter, drilling and milling operations are programmed using all three machine axes. The programs presented in this chapter are not what is referred to as true three-axis programming. The term *three-axis programming* is used to describe a program or program sequence in which *all three machine axes are used at the same time*. The type of programming presented in this chapter will use all three axes, but will primarily position a location using the X and Y axis first, then use the Z axis to perform a milling or drilling operations. This type of programming is called two-and-a-half axis programming. Two-and-a-half axis programming is the most common CNC mill programming done, accounting for about 90 percent of the CNC machining center work programmed.

Two-and-a-half axis programming also is the practical limit for manual programming. The manual mathematical calculations required for three-axis contouring are too time consuming to justify manual programming. True three-, four-, and five-axis programming are best done using a computer-aided system, such as a CAD/CAM system.

The programs in this chapter all use tool length offset compensation. The concept of tool length offset was discussed in Chapter 4. The programs presented in this chapter will put the concept to use. When setting up a job in a machine, the setup person, or operator, enters the tool lengths into the appropriate tool length offset registers in the CNC controller. Commands in the CNC program activate the tool length offset compensation feature for a given tool. The tool length compensation adjusts the Z-axis zero point to account for the differences in the lengths of the various cutting tools used in the program.

A PROGRAMMING TASK USING THREE AXES

The part in Figure 7-1 is to be milled. The program is to be written using absolute positioning in the word address format. The cutter path is given in Figure 7-2. The machine used to cut the part is a vertical machining center equipped with a FANUC 11M control.

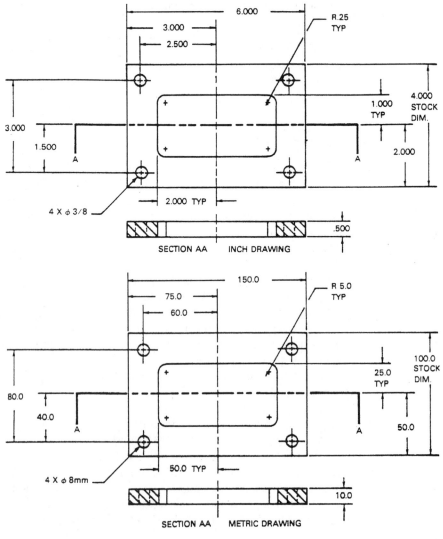

FIGURE 7-1
Part drawing for three-axis programming task

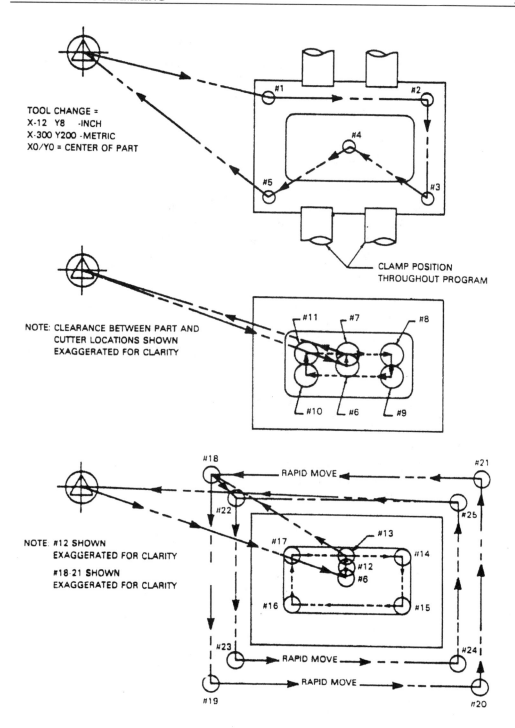

FIGURE 7-2

Cutter paths for part drawing in Figure 7-1

The following sequence of events is to be performed.

1. Move to the machine home zero (see discussion on home zero later in this chapter) and preset the part coordinate system.
2. Change tools, placing a drill into the spindle, and turn on the tool length offset compensation.
3. Drill hole #1.
4. Drill hole #2.
5. Drill hole #3.
6. Drill hole #4.
7. Drill hole #5.
8. Cancel the tool length offset compensation.
9. Change tools, placing a 1.000-inch diameter end mill in the spindle and turning on the tool length offset compensation.
10. Move to location 6 at rapid traverse, and plunge cut a hole through the part.
11. Feed from #6 to #7.
12. Feed from #7 to #8.
13. Feed from #8 to #9.
14. Feed from #9 to #10.
15. Feed from #10 to #11.
16. Feed from #11 to #7.
17. Retract the spindle and cancel the tool length offset compensation.
18. Change tools, placing a .500-inch diameter end mill in the spindle; turn on the tool length offset compensation.
19. Rapid traverse to location #6.
20. Lower the spindle to depth.
21. Feed from #6 to #13.
22. Feed from #13 to #14.
23. Feed from #14 to #15.
24. Feed from #15 to #16.
25. Feed from #16 to #17.
26. Feed from #17 to #13.
27. Feed from #13 to #12.
28. Retract the spindle
29. Rapid traverse from #12 to #18.
30. Lower the spindle to depth.
31. Feed from #18 to #19.
32. Retract the spindle.
33. Rapid traverse from #19 to #20, jumping over the clamps.
34. Lower spindle to depth.
35. Feed from #20 to #21.
36. Retract the spindle.
37. Rapid traverse from #21 to #18, jumping the clamps.
38. Lower spindle to depth.
39. Feed from #18 to #22.
40. Feed from #22 to #23.

41. Retract the spindle.
42. Rapid move from #23 to #24.
43. Lower spindle to depth.
44. Feed from #24 to #25.
45. Retract the spindle.
46. Turn off the tool length offset compensation.
47. Return the spindle to the home zero location.

The program manuscript is presented in Figure 7-3. The metric version is given in Figure 7-4. Several new word address commands are used in this program. They are:

G28 – Return to reference point command. G28 is used in conjunction with other commands to cause the spindle to position at the machine's coordinate system origin. This point is referred to as *home zero* in most CNC shops. If coordinates are specified on the G28 line, the spindle will first move to the coordinates, then to home zero. In this manner, the spindle may be moved to a known safe position before moving to home zero.

G44 – Calls up a tool length offset register. A G44 accomplishes a Z-zero shift toward the workpiece.

H – Used to assign a tool register. H01 would assign the information stored in tool length register #1. H02 would assign the information stored in tool length register #2.

G49 – This is the tool length offset cancel code.

G81 – This is the canned drill cycle. When a G81 is issued, the spindle rapids to the X/Y coordinates specified on the drill cycle line, the Z axis then rapids to the specified feed engagement point, feeds to the final drill depth, and then rapids out of the hole to either the rapid or initial level (as will be explained).

G80 – This is the canned cycle cancel code. When a G80 is issued, the active canned cycle code is turned off.

R – This address stands for the canned cycle reference level. The *reference level* is the spot where the programmer desires the canned cycle to start feeding into the workpiece. The reference level is also call the *rapid* or *gage* level.

G92 – Absolute zero set command. This command tells the control to reset the part coordinate system origin. Coordinates must be specified on the G92 block. The coordinates tell the machine where to set the origin, *relative to the current spindle position.*

G99/G98 – G98 is the return to initial level command. G99 is the return to rapid (reference) level command. When a canned cycle is active, the spindle may be directed to return to the rapid level when it exits a hole with a G99. If the programmer desires the spindle to return to the original starting point Z height, the G98 command is issued. G99 results in the faster cycle. G98 is particularly useful for jumping over clamps and other obstructions while in a cycle.

M01 – Program optional stop code. M01 functions as an M00 with one exception: it is only effective if the optional stop switch on the machine

```
%
07003
(* *********************************)
(* COORDINATE SYSTEM ORIGIN)
(* X/Y 0 - CENTERLINE OF PART)
(* Z0 - .100 ABOVE TOP OF PART)
(* *********************************)
(* ABSOLUTE ZERO SHIFT TO PART SYSTEM)
(* *********************************)
N10G80G90G70
N20G28G91X0.Y0.Z0.
N30G92X10.625Y7.5Z6.
(* ***********)
(* TOOL 1 - 3/8 STUB DRILL)
(* DRILL HOLES
(* ***********)
N40G00G90G98G70
N50T01M06
N60G00X-2.5Y1.5S1066M03T02
N70G44Z0.H01M08
N80G81G99X-2.5Y1.5Z-.62R0.F12.8
N90X2.5
N100Y-1.5
N110X0.Y0.
N120X-2.5Y-1.5
N130G80G00Z0.
N140G49M01
(*************)
(* TOOL 2 - 1.0 DIA. 4-FLT. END MILL
(* ROUGH MILL INSIDE OF SLOT
(* ***********)
N150G00G90G98G70
N160T02M06
N170G00X0.Y0.3S425M03T03
N180G44Z0.H02M08
(FEED TO DEPTH)
N190G01Z-.62F6.8
(ROUGH MILL INSIDE)
N200Y.48
N210X1.48
N220Y-.48
N230X-1.48
N240Y.48
N250X0.
N260G80G00Z0.M09
N270G49M01
(* ***********)
(* TOOL 3 - 1/2 DIA. 4-FLT. END MILL)
(* FINISH INSIDE SLOT)
(* ROUGH/FINISH OUTSIDE OF PART)
(* ***********)
N280G00G90G98G70
N280T03M06
N300G00X0.Y0.S800M03T03
```

FIGURE 7-3 *continued on next page*

```
N310G44Z0.H03M08
(FEED TO DEPTH)
N320G01Z-.62F12.8
(FINISH MILL INSIDE SLOT)
N330Y.75
N340X1.75
N350Y-.75
N360X-1.75
N370Y.75
N380X0.
(PULL AWAY FROM PART AND RETRACT SPINDLE)
N390Y.74
N400G00Z3.
(POSITION TO START OF OUTSIDE MILL CUT)
N410X-3.26Y2.26
N420Z0.
(FEED TO DEPTH AND ROUGH MILL 1ST SIDE)
N430G01Z-.62F12.8
N440Y-2.26
(RETRACT SPINDLE AND JUMP OVER CLAMP)
(POSITION FOR ROUGH CUT ON 2ND SIDE)
N450G00Z3.
N460X3.26
N470Z0.
(FEED TO DEPTH AND ROUGH MILL 2ND SIDE)
N480G01Z-.62F12.8
N490Y2.26
(RETRACT SPINDLE AND JUMP OVER CLAMP)
N500G00Z3.
N510X-3.26Y2.26
N520Z0.
(FEED TO DEPTH - MOVE TO PART SURFACE)
(AND FINISH MILL FIRST SIDE)
N530G01Z-6.2F12.8
N540X-3.25Y2.25
N550Y-2.25
(RETRACT SPINDLE AND JUMP OVER CLAMP)
(POSITION FOR FINISH CUT ON 2ND SIDE)
N560G00Z3.
N570X3.25
N580Z0.
(FEED TO DEPTH AND FINISH MILL 2ND SIDE)
N590G01Z-.62F12.8
N600Y2.25
N610G80G00Z0M09
N620G49
N630G28G91Z0.M05
N640G28X0.Y0.
N650M30
%
```

FIGURE 7-3

Three-axis program, inch, for the part in Figure 7-1

```
%
07004
(* **************************)
(* COORDINATE SYSTEM ORIGIN)
(* X/YO - CENTERLINE OF PART)
(* ZO - .100 ABOVE TOP OF PART)
(* **************************)
(* ABSOLUTE ZERO SHIFT TO PART SYSTEM)
(* **************************)
N10G80G90G71
N20G28G91X0.Y0.Z0.
N30G92X269.875Y190.Z152.4
(* ***********)
(* TOOL 1 - 3/8 STUBB DRILL)
(* DRILL HOLES
(* ***********)
N40G00G90G98G70
N50T01M06
N60G00X-60.Y40.S1066M03T02
N70G44Z0.H01M08
N80G81G99X-60.Y40.Z-13.R0.F216.
N90X60.
N100Y-40.
N110X0.Y0.
N120X-60.Y-40.
N130G80G00Z0.
N140G49M01
(* ***********)
(* TOOL 2 - 1.0 DIA. 4-FLT. END MILL
(* ROUGH MILL INSIDE OF SLOT
(* ***********)
N150G00G90G98G70
N160G02M06
N170G00X0.Y0.S425M03T03
N180G44Z0.H02M08
(FEED TO DEPTH)
N190G01Z-13.F172.5
(ROUGH MILL INSIDE)
N200Y12.25
N210X37.25
N220Y-12.25
N230X-37.25
N240Y12.25
N250X0.
N260G80G00Z0.M09
N270G49M01
(* ***********)
(* TOOL 3 - 1/2 DIA. 4-FLT. END MILL)
(* FINISH INSIDE SLOT).
(* ROUGH/FINISH OUTSIDE OF PART)
(* ***********)
N280G00G90G98G70
N290T03M06
N300G00X0.Y0.S800M03T01
```

FIGURE 7-4 continued on next page

```
N310G44Z0.H03M08
(FEED TO DEPTH)
N320G01Z-13.F325.1
(FINISH MILL INSIDE SLOT)
N330Y20.
N340X45.
N350Y-20.
N360X-45.
N370Y20.
N380X0.
(PULL AWAY FROM PART AND RETRACT SPINDLE)
N390Y44.75
N400G00Z75.
(POSITION TO START OF OUTSIDE MILL CUT)
N410X-80.25Y55.25
N420Z0.
(FEED TO DEPTH AND ROUGH MILL 1ST SIDE)
N430G01Z-13.F325.1
N440Y-55.25
(RETRACT SPINDLE AND JUMP OVER CLAMP)
(POSITION FOR ROUGH CUT ON 2ND SIDE)
N450G00Z75.
N460X80.25
N470Z0.
(FEED TO DEPTH AND ROUGH MILL 2ND SIDE)
N480G01Z-13.325.1
N490Y55.25
(RETRACT SPINDLE AND JUMP OVER CLAMP)
N500G00Z75.
N510X-80.25
N520Z0.
(FEED TO DEPTH - MOVE TO PART SURFACE)
(AND FINISH MILL FIRST SIDE)
N530G01Z-13.F325.1
N540X-80.Y55.
N550Y-55.
(RETRACT SPINDLE AND JUMP OVER CLAMP)
(POSITION FOR FINISH CUT ON 2ND SIDE)
N560G00Z75.
N570X80.
N580Z0.
(FEED TO DEPTH AND FINISH MILL 2ND SIDE)
N590G01Z-13.F325.1
N600Y55.
N610G80G00Z0M09
N620G49
N630G28G91Z0.M05
N640G28X0.Y0.
N650M30
%
```

FIGURE 7-4

Three-axis program, metric, for the part in Figure 7-1

control is turned on. When this switch, called an opstop switch, is off, the M01 is ignored by the control.

M03 – M functions, as briefly explained in Chapter 6, control a number of auxiliary functions. M03 is the code for turning the spindle on in the clockwise direction.

M05 – Turns the spindle off.

M06 – Tool change code. When M06 is issued, the machine's automatic tool changer sequence will be initiated.

M08 – Turns the flood coolant on.

M09 – Turns the coolant off.

T – Selects the tool to be put in the spindle by the tool changer.

F – Assigns feedrates, as in two-axis programming.

S – Designates the spindle speed.

Modal/Nonmodal Commands

In the program in Figure 7-3, as well as in Chapter 6, certain commands remain active until cancelled by another code. Codes that are active for more than the line in which they are issued are called *modal* commands. Rapid traverse, feedrate moves, and canned cycle codes are all examples of modal commands. A *nonmodal* command is one that is active only in the program block in which it is issued. M00 program stop is an example of a nonmodal command.

Canned Cycles

Canned cycles, such as the G81 codes explained previously, are routines, built into the control, to perform standard operations. Drilling, boring, and tapping, for instance, are common operations. Rather than the programmer having to program each move associated with perfoming one of these operations, he/she can call a canned cycle instead. The following line institutes a canned cycle for tapping.

G84G99X1.Y.375Z-.753R.1F10.

The codes used are

G84 – G-code to turn on the tapping cycle. The spindle will feed into the workpiece with the spindle rotating clockwise, stop at the programmed Z axis coordinate, reverse the spindle, then feed back out of the workpiece until it reaches the programmed feed engagement point.

G99 – Specifies that the spindle should return to the reference level (the feed engagement point) when retracting out of the hole.

X/Y coordinates – Indicate the location where the cycle is to begin. The spindle will first position here at rapid traverse before moving the Z-axis.

Z coordinate – Tells the control how deep to feed the Z-axis. It is the actual Z coordinate to which the spindle is to move.

R – Specifies the Z coordinate where the spindle is to begin feeding. Until the spindle reaches this coordinate, it will move in rapid traverse.

F – Sets the feedrate for the Z-axis feedrate moves.

A complete cycle would look like the following:

G84G99X1.Y.375Z-.753R.1F10.
X1.5
Y1.375
X1.
G80

Each coordinate causes the spindle to position at that coordinate and repeat the cycle. The G80 turns off the cycle. There are a number of canned cycles available on CNC controls.

PROGRAM EXPLANATION (Refer to Figures 7-3)

%

Program start code.

O7003

Program number.

N10 – N30 are the tape startup blocks
N10

N10 – The block sequence number.
G80 – Canned cycle cancel command, turns off any active canned cycles.
G90 – Selects absolute positioning mode.
G70 – Selects inch input.

N20

N20 – The block sequence number.
G28 – Return to reference point. On FANUC-style controls, the reference point is the machine home zero. The machine is returned to home zero prior to issuing a G92 coordinate system preset in the next block. Even if the operator moved the spindle between program cycles, the spindle would be positioned to home zero because this block was included in the program.
G91 – Selects absolute positioning.
X0.Y0. coordinates – Because the machine is in incremental mode, the spindle will not move anywhere. The G28 command simply will return the spindle to home zero.

N30

N30 – The block sequence number.
G92 – Absolute zero preset code. The G92 cause the control to reset the part coordinate system. The G28 command in the previous block insured the spindle was at home zero prior to issuing the G92.
X/Y coordinates – Specify where the part origin should be set, incrementally from the current spindle position.

N40 – *Tool sequence safety block*

N40 – The block sequence number.

G00 – Selects rapid traverse mode.

G90 – Selects absolute positioning.

G98 – Selects return to initial level mode on canned cycle Z-axis retract moves.

G70 – Selects inch input.

N50 through N70 are the tool change blocks
N50

N50 – The block sequence number.

T01 – Places tool 1 in standby mode for the next tool change command.

M06 – Initiates an automatic tool change cycle. The tool in standby (in this case tool 1) is placed in the spindle, and the previous tool put away in the tool storage magazine.

N60

N60 – The block sequence number.

G00 – Selects rapid traverse mode.

X/Y coordinates – Move the spindle from home zero to the position #1.

S1066 – Sets the spindle speed to 1066 RPM.

M03 – Turns the spindle on clockwise.

T02 – Places tool 2 in tool change standby.

N70

N70 – The block sequence number.

G44 – Turns on tool length compensation.

Z coordinate – Moves the Z-axis to .100 above the top of the part (the part Z-zero point).

H01 – Instructs the control to use the values in tool length offset register #1 for tool length compensation.

M08 – Turns on the flood coolant.

N80 through N120 are the tool motion sequence
N80

N80 – The block sequence number.

G81 – Turns on the canned drilling cycle.

G99 – Instructs the control to return the Z-axis to the feed engagement point (the reference level) when retracting the spindle out of a hole. The G99 is effective only when the canned cycle is turned on.

X/Y coordinates – Coordinates of hole #1.

Z coordinate – The depth of the drilled hole. Note that this is the programmed depth (i.e., where the tip of the tool is to be sent).

R – Specifies the Z-axis coordinate for the feed engagement point. This point is also referred to as the r-plane. In this example, the spindle is commanded to begin feeding into the workpiece at the Z-zero point (.100 above the top of the part).

F12.8 – Sets feedrate at 12.8 inches per minute.

N90

N90 – The block sequence number.

X coordinate – Moves the spindle from position #1 to #2.

N100

N100 – The block sequence number.

Y coordinate – Moves the spindle from position #2 to #3.

N110

N110 – The block sequence number.

X/Y coordinate – Moves the spindle from position #3 to #4.

N120

N120 – The block sequence number.

X/Y coordinate – Moves the spindle from position #4 to #5.

N130 and N140 are the tool cancel sequence
N130

N130 – The block sequence number.

G80 – Turns off the canned drill cycle.

G00 – Selects rapid traverse mode.

Z0. – Returns the spindle to the part Z-zero point.

N140

N140 – The block sequence number.

G49 – Cancels the tool length compensation.

M01 – Optional program stop code. The M01 is included for operator convenience. If the operator desired to check the part or setup for any reason, he or she need only turn on the opstop switch on the control. The program will halt execution at the end of the tool cycle.

N150 – *Tool sequence safety block*

N150 – The block sequence number.

G00 – Selects rapid traverse mode.

G90 – Selects absolute positioning.

G98 – Selects return to initial level mode on canned cycle Z-axis retract moves.

G70 – Selects inch input.

N160 through N180 are the tool change blocks
N160

N160 – The block sequence number.

T02 – Places tool 2 in standby mode for the next tool change command.

M06 – Initiates an automatic tool change cycle.

N170

N170 – The block sequence number.

G00 – Selects rapid traverse mode.

X/Y coordinates – Move the spindle from home zero to the position #6.

S – Sets the spindle speed rpm.

M03 – Turns the spindle on clockwise.

T03 – Places tool 3 in tool change standby.

N180

N180 – The block sequence number.

G44 – Turns on tool length compensation.

Z coordinate – Moves the Z-axis to .100 above the top of the part (the part Z-zero point).

H02 – Instructs the control to use the values in tool length offset register #2 for tool length compensation.

M08 – Turns on the flood coolant.

N190 through N250 are the tool motion sequence
N190

N190 – Block sequence number.

G01 – Selects feedrate mode.

Z coordinate – Positions the spindle, at feedrate, to the milling depth.

F – Sets the feedrate.

N200

N200 – Block sequence number.

Y coordinate – Moves the spindle from position #6 to #7.

N210

N210 – Block sequence number.

X coordinate – Moves the spindle from position #7 to #8.

N220

N220 – Block sequence number.

Y coordinate – Moves the spindle from position #8 to #9.

N230

N230 – Block sequence number.

X coordinate – Moves the spindle from position #9 to #10.

N240

N240 – Block sequence number.

Y coordinate – Moves the spindle from position #10 to #11.

N250

N250 – Block sequence number.

X coordinate – Moves the spindle from position #11 to #7.

N260 & N270 are the tool cancel sequence
N260

N260 – The block sequence number.

G80 – Turns off the canned drill cycle.

G00 – Selects rapid traverse mode.

Z0. – Returns the spindle to the part Z-zero point.

N270

N140 – The block sequence number.

G49 – Cancels the tool length compensation.

M01 – Optional program stop code.

N280 – *Tool sequence safety block*

N280 – The block sequence number.

G00 – Selects rapid traverse mode.

G90 – Selects absolute positioning.

G98 – Selects return to initial level mode on canned cycle Z-axis retract moves.

G70 – Selects inch input.

N290 through N310 are the tool change blocks
N290

N290 – The block sequence number.

T03 – Places tool 3 in standby mode for the next tool change command.

M06 – Initiates an automatic tool change cycle.

N300

N300 – The block sequence number.

G00 – Selects rapid traverse mode.

X/Y coordinates – Move the spindle from home zero to the position #6.

S – Sets the spindle speed rpm.

M03 – Turns the spindle on clockwise.

T01 – Places tool 1 in tool change standby.

N310

N310 – The block sequence number.

G44 – Turns on tool length compensation.

Z coordinate – Moves the Z-axis to .100 above the top of the part (the part Z-zero point).

H03 – Instructs the control to use the values in tool length offset register #3 for tool length compensation.

M08 – Turns on the flood coolant.

N320 through N600 are the tool motion sequence
N320

N320 – Block sequence number.

G01 – Selects feedrate mode.

Z coordinate – Positions the spindle, at feedrate, to the milling depth.

F – Sets the feedrate.

N330

N330 – Block sequence number.

Y coordinate – Moves the spindle from position #6 to #13.

N340
>   N340 – Block sequence number.
>   X coordinate – Moves the spindle from position #13 to #14.

N350
>   N350 – Block sequence number.
>   Y coordinate – Moves the spindle from position #14 to #15.

N360
>   N360 – Block sequence number.
>   X coordinate – Moves the spindle from position #15 to #16.

N370
>   N370 – Block sequence number.
>   Y coordinate – Moves the spindle from position #16 to #17.

N380
>   N380 – Block sequence number.
>   X coordinate – Moves the spindle from position #17 to #13.

N390
>   N390 – Block sequence number.
>   X coordinate – Moves the spindle from position #13 to #12.

N400
>   N400 – Block sequence number.
>   G00 – Selects rapid traverse mode.
>   Z coordinate – Retracts spindle to Z3.000.

N410
>   N410 – Block sequence number.
>   X/Y coordinates – Move the spindle from position #12 to #18.

N420
>   N420 – Block sequence number.
>   Z coordinate – Rapids spindle to .100 above the part.

N430
>   N430 – Block sequence number.
>   G01 – Selects feedrate mode.
>   Z coordinate – Moves the spindle at feedrate to the milling depth.
>   F – Sets the feedrate.

N440
>   N440 – Block sequence number.
>   Y coordinate – Moves the spindle from position #18 to #19.

N450
>   N450 – Block sequence number.
>   G00 – Selects rapid traverse mode.
>   Z coordinate – Retracts spindle to Z3.000.

N460

N460 – Block sequence number.
X/Y coordinates – Move the spindle from position #19 to #20.

N470

N470 – Block sequence number.
Z coordinate – Rapids spindle to .100 above the part.

N480

N480 – Block sequence number.
G01 – Selects feedrate mode.
Z coordinate – Moves the spindle at feedrate to the milling depth.
F – Sets the feedrate.

N490

N490 – Block sequence number.
Y coordinate – Moves spindle from position #20 to #21.

N500

N500 – Block sequence number.
G00 – Selects rapid traverse mode.
Z coordinate – Retracts spindle to Z3.000.

N510

N510 – Block sequence number.
X/Y coordinates – Move the spindle from position #21 to #18.

N520

N520 – Block sequence number.
Z coordinate – Rapids spindle to .100 above the part.

N530

N530 – Block sequence number.
G01 – Selects feedrate mode.
Z coordinate – Moves the spindle at feedrate to the milling depth.
F – Sets the feedrate.

N540

N540 – Block sequence number.
X/Y coordinates – Moves spindle from position #18 to #22.

N550

N550 – Block sequence number.
Y coordinate – Moves spindle from position #22 TO #23.

N560

N560 – Block sequence number.
G00 – Selects rapid traverse mode.
Z coordinate – Retracts spindle to Z3.000.

N570

N570 – Block sequence number.

X/Y coordinates – Move the spindle from position #23 to #24.

N580

N580 – Block sequence number.

Z coordinate – Rapids spindle to .100 above the part.

N590

N590 – Block sequence number.

G01 – Selects feedrate mode.

Z coordinate – Moves the spindle at feedrate to the milling depth.

F – Sets the feedrate.

N600

N600 – Block sequence number.

Y coordinate – Moves spindle from position #24 to #25.

N610 and N620 are the tool cancel sequence
N610

N610 – The block sequence number.

G80 – Turns off the canned drill cycle.

G00 – Selects rapid traverse mode.

Z0. – Returns the spindle to the part Z-zero point.

M09 – Turns off the flood coolant.

N620

N620 – The block sequence number.

G49 – Cancels the tool length compensation.

N630 through N650 are the end of tape sequence
N630

N630 – Block sequence number.

G28 – Return to reference code.

G91 – Selects incremental positioning.

Z0. – In conjunction with G28G91 commands the Z-axis to the home zero position.

M05 – Turns off the spindle.

N640

N640 – Block sequence number.

G28 – Return to reference code.

X0.Y0. – In conjunction with G28 commands the X and Y axis to the home zero position.

N650

N650 – Block sequence number.

M30 – Signals end of program.

%

Program stop code.

FIGURE 7-5
A light-duty vertical machining center *(Photo courtesy of Bridgeport Machines Division of Textron Inc.)*

OTHER G CODES USED IN CNC PROGRAMMING

In the examples just presented, the basic drill cycle G81 was used. There are a number of other G codes that can be used in CNC program. A list of these is included in Appendix 3. Some of the more common canned cycles are diagrammed in Figures 7-6, 7-7, and 7-8.

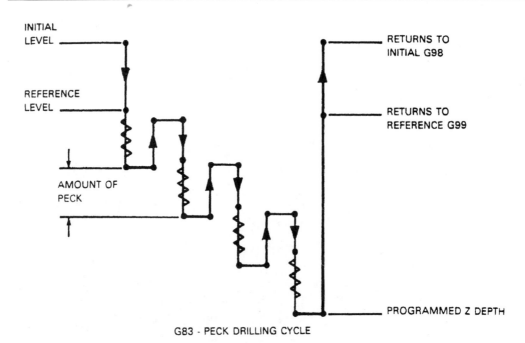

INITIAL LEVEL

REFERENCE LEVEL

AMOUNT OF PECK

RETURNS TO INITIAL G98

RETURNS TO REFERENCE G99

PROGRAMMED Z DEPTH

G83 - PECK DRILLING CYCLE

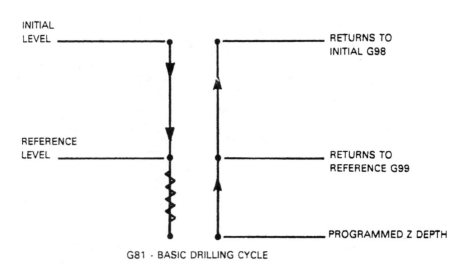

INITIAL LEVEL

REFERENCE LEVEL

RETURNS TO INITIAL G98

RETURNS TO REFERENCE G99

PROGRAMMED Z DEPTH

G81 - BASIC DRILLING CYCLE

———————— = RAPID MOVEMENT   ∿∿∿∿ = FEEDRATE MOVEMENT

FIGURE 7-6
G codes for peck drilling and basic drilling cycles

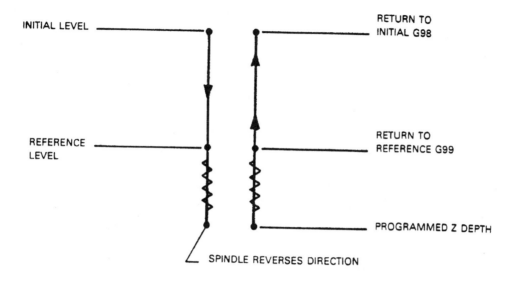

G84 - TAPPING CYCLE

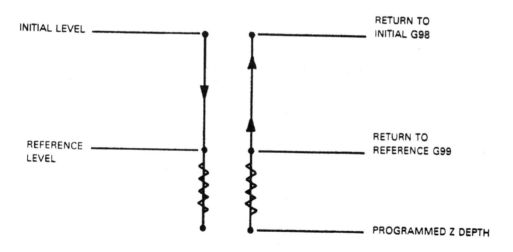

G85 - BORING CYCLE, TYPE A

= RAPID MOVEMENT   = FEEDRATE MOVEMENT

FIGURE 7-7
G codes for tapping and boring cycles

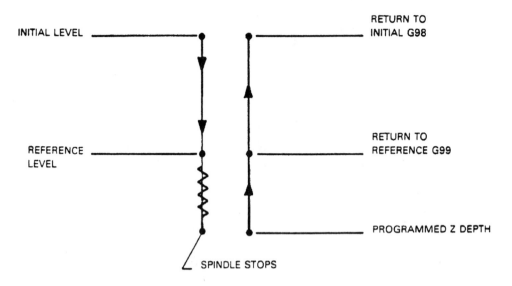

INITIAL LEVEL

RETURN TO
INITIAL G98

REFERENCE
LEVEL

RETURN TO
REFERENCE G99

PROGRAMMED Z DEPTH

SPINDLE STOPS

G86 BORING CYCLE TYPE B

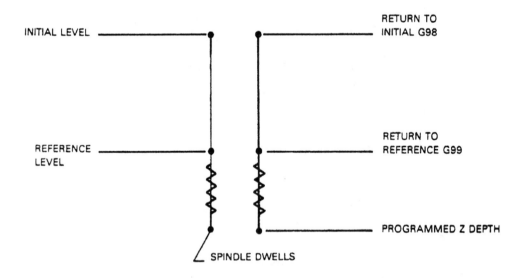

INITIAL LEVEL

RETURN TO
INITIAL G98

REFERENCE
LEVEL

RETURN TO
REFERENCE G99

PROGRAMMED Z DEPTH

SPINDLE DWELLS

G89 - BORING CYCLE, TYPE C

——————— = RAPID MOVEMENT   ⋀⋀⋀⋀ = FEEDRATE MOVEMENT

FIGURE 7-8
G codes for boring cycles

USING AN INDEXER

Many CNC machining centers use a fourth axis in the form of a rotary table or indexer. This concept was discussed in Chapter 2. The common rotational axis are

A-axis = rotation around the X-axis.
B-axis = rotation around the Y-axis.
C-axis = rotation around the Z-axis.

Often this fourth axis is used only to orient the part at a specific angle. The CNC programming used on the machine remains 2-1/2 or 3-axis in these cases. The part in Figure 7-9 will utilize a fourth axis index to help position the part.

Indexers vs. Rotary Tables

In the shop the terms *indexer* and *rotary table* are often used interchangeably. However, there is a specific difference between the two. A rotary table is a true fourth machine axis. It can rotate to any degree or fraction of a degree. An indexer, on the other hand, can only rotate in specific increments (usually 1 degree increments). Thus, an index can position a part for machining, while a rotary table can be used to perform contour milling. Machining centers with rotary tables are called four-axis machines; those with indexers are called 3-1/2-axis machines.

PROGRAMMING EXAMPLE

The part in Figure 7-9 is to be machined on a CNC vertical machining center using a FANUC 6M control. This particular setup will utilize two fixtures on the same machine. The first fixture is a set of vise jaws. Most of the part will be machined in this first fixture. To hold the part vertically to allow milling of the .625 wide × .125 deep slot on the end, a simple fixture will be mounted on the machine's indexer. The indexer will serve only to hold the part in the proper orientation. The programming used will still be 2-1/2 axis.

The indexer used on this machine has the capability of rotating in 1-degree increments. It is constructed so the rotation occurs around the X-axis, thus this axis is the A-axis. The indexer coordinates are specified in degrees. To command the indexer to move to 225 degrees, the command A225. is given in the tape. Indexers are simple to program.

The cutter path for the part in Figure 7-9 is given in Figures 7-10 and 7-11. The locations numbered on the cutter path are used to describe both the roughing and finishing passes. The roughing passes will leave .010 stock on the part for the finish passes. Figure 7-12 is the program manuscript. The setup sheet for the part is shown in Figure 7-13. The part coordinate system origin for this particular part is in the upper left corner. The Z-zero point is 1.000 inch above the part.

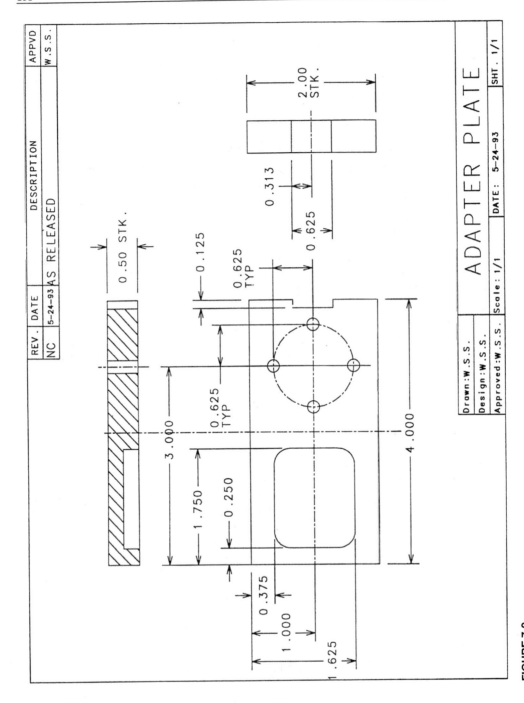

FIGURE 7-9
Part drawing for three-axis program using an indexer

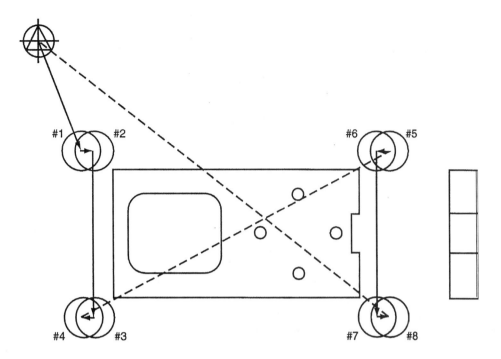

FIGURE 7-10
Cutter path for milling the outside of the part in Figure 7-9

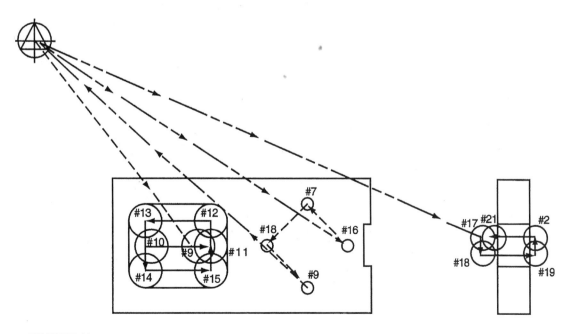

FIGURE 7-11
Cutter path for milling the slots and drilling the holes in the part in Figure 7-9

```
%
O7012
(* **********************************)
(* ADAPTER PLATE OPERATION 10 TAPE 7012)
(* **********************************)
N1G07G08G04X1.Y2.Z3.
N2G92X-12.752Y-7.453Z.0
(* **********)
(* TOOL NO. 1)
(* .615 DIA. END MILL)
(* ROUGH/FINISH 4.000 DIM.)
(* **********)
N3G40G90G00G80T01
N4G00X-.5Y.1S1200M03T02
N5G44Z-.75H01
N6G01X-.3225F7.2
N7Y-2.1
N8G00X-.5
N9Y.1
N10G01X-.3125
N11Y-2.1
N12G00Y.1
N13Z.0
N14X4.5Y-2.1
N15Z-1.05
N16G01X4.3225
N17Y.1
N18G00X4.5
N19Y-2.1
N20G01X4.3125
N21Y.1
N22G00X4.5
N23Z.0
N24G00G80Z.0
N25G49
N26M01
(* **********)
(* TOOL NO. 2)
(* .500 DIA. END MILL)
(* ROUGH/FINISH MILL POCKET)
(* **********)
N27G40G90G00G80T02
N28G00X1.33Y-1.S1500M03T03
N29G44Z.0H02
N30G01X.67Z-.74F9.
N31X1.49
N32Y-.635
N33X.51
N34Y-1.365
N35X1.49
N36Y-1.
N37X1.33
N38Z-.75
N39X1.5
```

FIGURE 7-12 *continued on next page*

```
N40Y-.625
N41X.5
N42Y-1.375
N43X1.5
N44Y-1.
N45Z-.753F.05
N46X.67
N47G00G80Z.0
N48G49
N49M01
(* **********)
(* TOOL NO. 3)
(* .188 DIA. STUB DRILL)
(* DRILL HOLE PATTERN)
(* **********)
N50G40G90G00G80T03
N51G00X3.Y.0S1800M03
N52G44Z.0H03
N53G83G99X3.625Y.0Z-1.1064R-.4F7.2
N54X3.Y.625
N55X2.375Y.0
N56X3.Y-.625
N57G80
N58G00G80Z.0
N59G49
N60M01
N61G91G28X.0Y.0Z.0
N62M30
%
```

FIGURE 7-12
Program for part in Figure 7-9

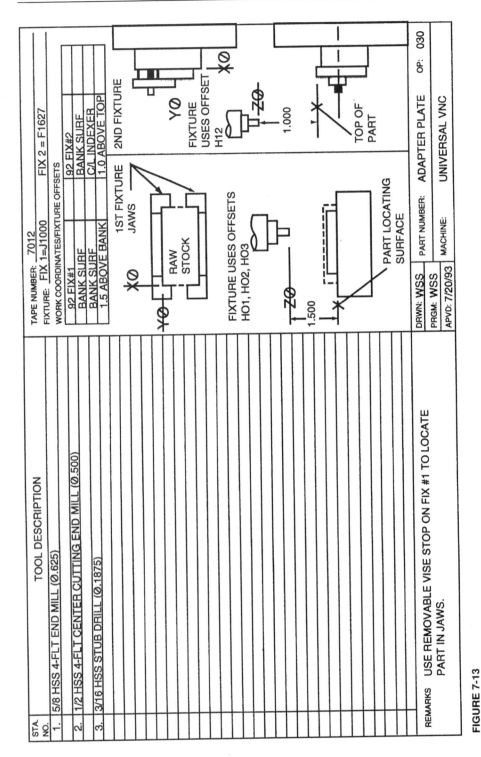

FIGURE 7-13
Setup sheet for part in Figure 7-9

Program Explanation

N1
Send spindle to home zero position.

N2
Set the part coordinate system for the second fixture.

N3 through N5
Place tool 1 in spindle, put tool 2 in tool change standby, position spindle at location #1, and turn on tool length compensation.

N6
Move spindle from position #1 to #2 at feedrate, positioning the spindle to begin a rough pass.

N7
Move spindle from position #2 to #3 at feedrate.

N8
Move spindle away from the part surface, from #3 to #4 at rapid.

N9
Rapid spindle from position #4 back to #1.

N10
Feed spindle from position #1 to #2, positioning the spindle for the start of a finish pass.

N11
Feed the spindle from #2 to #3, completing the finish pass.

N12
Rapid the spindle away from the part surface, from #3 to #4.

N13
Retract the spindle at rapid above the part.

N14
Rapid from position #4 to #5.

N15
Rapid the Z-axis to the final milling depth.

N16
Move spindle from position #5 to #6 at feedrate, positioning the spindle to begin a rough pass.

N17
Move spindle from position #6 to #7 at feedrate

N18
Move spindle away from the part surface, from #7 to #8 at rapid.

N19

Rapid spindle from position #8 back to #5.

N20

Feed spindle from position #5 to #6, positioning the spindle for the start of a finish pass.

N21

Feed the spindle from #6 to #7, completing the finish pass.

N22

Rapid the spindle away from the part surface, from #7 to #8.

N23

Retract the spindle at rapid above the part.

N24 through N26

Turn off the tool length compensation, and issue an M01 program optional stop code for operator convenience.

N27 through N29

Place tool 2 in spindle, tool 3 in standby. Position to location #9, and turn on tool length compensation.

N30

Move spindle from position #9 to #10 at feedrate.

Notice that this move was made in both X and Z. In this manner the cutter is ramped into the part rather than plunged. This requires much less cutting tool pressure. Although on a shallow pocket such as this one a straight plunge would work effectively, on deep pockets this ramping move prevents tool breakage.

N31

Move spindle from position #10 to #11 at feedrate. Not only does this rough the center area of the pocket, but it also removes the ramp left from the previous block.

N32

Move spindle from #11 to #12 at feedrate.

N33

Move spindle from #12 to #13.

N34

Move spindle from #13 to #14.

N35

Move spindle from #14 to #15.

N36

Move spindle from #15 to #11.

N37

Move spindle form #11 to #10.

N38

Position Z-axis to final milling depth.

N39

Move spindle from position #10 to #11.

N40

Move spindle from #11 to #12.

N41

Move spindle from #12 to #13.

N42

Move spindle from #13 to #14.

N43

Move spindle from #14 to #15.

N44

Move spindle from #15 to #11.

N45

Move spindle away from the part, .003 in X, .100 in Y. This small X-axis departure prevents the cutter from leaving a mark on the side of the pocket when the cutter stops its motion, prior to retracting the Z-axis.

N46 through N48

Tool cancel blocks.

N49 through N51.

Place tool 3 in the spindle, tool 2 in standby. Turn on the tool length compensation, and position spindle over hole location #16.

N52

Turn on canned drill cycle and drill hole #16.

N53 through N55

Drill holes #17, 18, 19, and 20.

N66

Cancel the canned drill cycle.

N57 through N59

Tool cancel blocks.

N60 through 61

Return spindle to home zero. Establish a new part coordinate system for the second fixture.

N62

Put tool 2 in the spindle.

N63

Rotate the indexer (the A-axis) to zero degrees. This orients the part with the end to be machined up.

N64 through N65

Move to position #17, and turn on the tool length compensation. Notice that a different offset register is used for tool 2 this time. The original tool length offset was set relative to the first fixture's part coordinate system. It was necessary, therefore, to use a different register for fixture 2.

N66

Move from position #17 to #18. This positions the spindle for a roughing pass. There is .010 stock left on the sides of the slot at this point.

N67

Move from position #18 to #19.

N68

Move from #19 to #20.

N69

Move from #20 to #17.

N70

Move from #17 to #18. This positions the spindle for the finish pass.

N71

Move from #18 to #19.

N72

Move from #19 to #20.

N73 through N75

Tool cancel blocks.

N76 through N78

End of program blocks.

SUMMARY

- The important concepts presented in this chapter are: Tool lengths in three-axis machines must be preset by the operator. On some controls they can be preset in the program.
- The initial level is the Z-axis spindle position when an 80 series canned cycle commences. A reference (or rapid) level is the Z-axis feedrate engagement point, selected by the programmer. G98 selects a return to initial level, and G99 selects a return to reference level when using 80 series G codes (canned cycle codes).

- Canned cycles are routines built into the controller to simplify programming. Values, called parameters, are passed to the control indicating how the cycle is to perform, where the cycle is to begin, and how it should repeat.
- Positioning the spindle in two axis, then feeding with the third is called 2-1/2-axis programming. Feeding with all three axes simultaneously is called 3-axis programming.
- Indexers often are used on CNC machinery. Positioning an index usually is just a matter of calling out the axis designator and a coordinate (i.e., A0., B270., A135.).

VOCABULARY INTRODUCED IN THIS CHAPTER

2-1/2-axis programming
3-axis programming
4-axis programming
Canned cycle
Indexer
Initial level
Rapid level
Reference level
Rotary table

REVIEW QUESTIONS

1. What is the difference between 3-axis programming and 2-1/2-axis programming?
2. How are tool lengths called up? How are they cancelled?
3. What is a canned cycle?
4. What is a feed engagement point in a canned cycle?
   What is the reference level?
   What is the initial level?
5. What is the difference between a G98 and G99 in a canned cycle?
6. What is the difference between a modal and nonmodal command?
7. How is absolute positioning specified in a program?
   How is incremental positioning specified?
8. What is an indexer?
9. How does an indexer differ from a rotary table?
10. What axis words do indexers and rotary tables use?

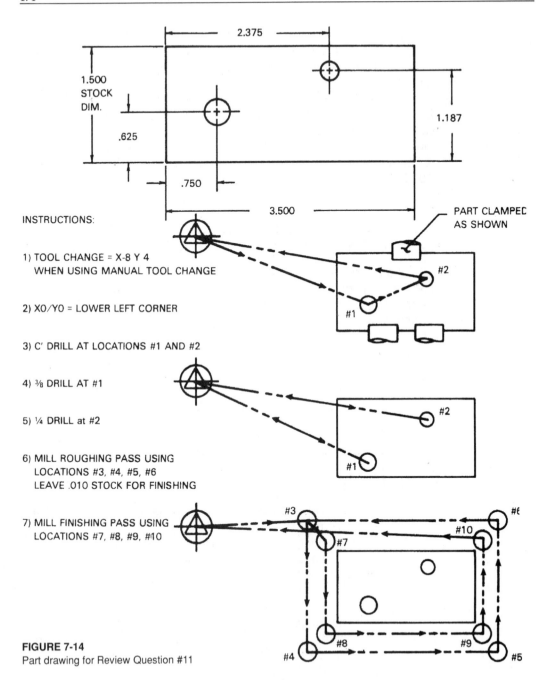

INSTRUCTIONS:

1) TOOL CHANGE = X-8 Y 4
   WHEN USING MANUAL TOOL CHANGE

2) X0/Y0 = LOWER LEFT CORNER

3) C' DRILL AT LOCATIONS #1 AND #2

4) ⅜ DRILL AT #1

5) ¼ DRILL at #2

6) MILL ROUGHING PASS USING
   LOCATIONS #3, #4, #5, #6
   LEAVE .010 STOCK FOR FINISHING

7) MILL FINISHING PASS USING
   LOCATIONS #7, #8, #9, #10

FIGURE 7-14
Part drawing for Review Question #11

11.   Write a program to mill and drill the part in Figure 7-14.

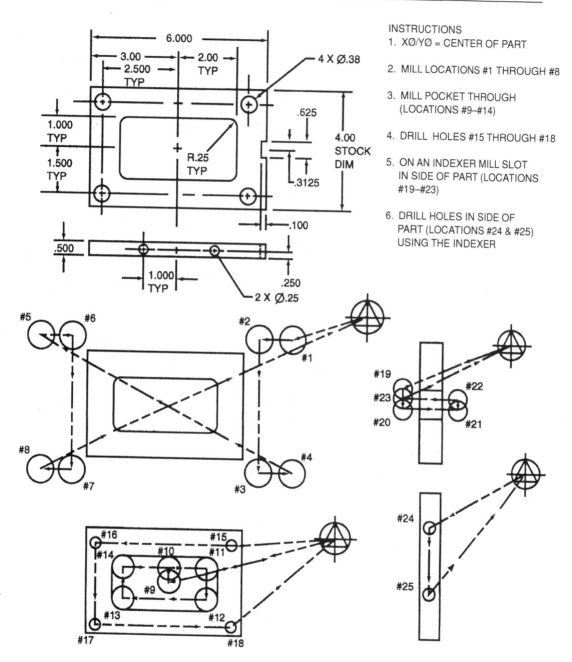

INSTRUCTIONS
1. XØ/YØ = CENTER OF PART

2. MILL LOCATIONS #1 THROUGH #8

3. MILL POCKET THROUGH
   (LOCATIONS #9–#14)

4. DRILL HOLES #15 THROUGH #18

5. ON AN INDEXER MILL SLOT
   IN SIDE OF PART (LOCATIONS
   #19–#23)

6. DRILL HOLES IN SIDE OF
   PART (LOCATIONS #24 & #25)
   USING THE INDEXER

FIGURE 7-15
Part drawing for Review Question #12

12. Write a program to mill and drill the part in Figure 7-15.

CHAPTER 8

Math for Numerical Control Programming

OBJECTIVE Upon completion of this chapter, you will be able to:

- Use right-angle trigonometry to determine programming coordinates from part drawings.

In the following chapters, the machining of arcs and angles will be discussed. For students already possessing a good working knowledge of trigonometry, this chapter will serve as a review. It is included here for students who have either not taken a course in shop math or who feel a review is in order.

What the machinist is able to do by blending arcs and angles through skill and feel for the craft, the NC part programmer must put into numeric coordinates. It is necessary for the programmer to become proficient at trigonometry to accomplish this task. Trigonometry has applications not only in NC programming but also in other types of machining situations. It is easily mastered with a little practical experience.

BASIC TRIGONOMETRY

Trigonometry is the mathematical science dealing with the solution of triangles. For example, knowing one side plus one other angle or side of a right triangle, all other information concerning the triangle can be derived using trigonometry. For machine stop use, the types of triangles usually dealt with are right triangles (see Figure 8-1). Note that one of the angles in the triangle is 90 degrees. A 90-degree angle is called a right angle: hence the name right triangle. The following formulas are also given in Figure 8-1:

$$\text{SINE} = \frac{\text{OPPOSITE SIDE}}{\text{HYPOTENUSE}}$$

$$\text{COSINE} = \frac{\text{SIDE ADJACENT}}{\text{HYPOTENUSE}}$$

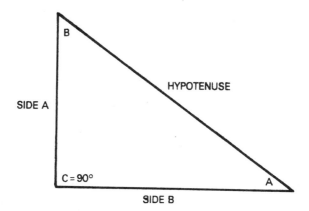

SINE $= \dfrac{\text{OPPOSITE SIDE}}{\text{HYPOTENUSE}}$ COSECANT $= \dfrac{\text{HYPOTENUSE}}{\text{SIDE OPPOSITE}}$

COSINE $= \dfrac{\text{SIDE ADJACENT}}{\text{HYPOTENUSE}}$ SECANT $= \dfrac{\text{HYPOTENUSE}}{\text{SIDE ADJACENT}}$

TANGENT $= \dfrac{\text{SIDE OPPOSITE}}{\text{SIDE ADJACENT}}$ COTANGENT $= \dfrac{\text{SIDE ADJACENT}}{\text{SIDE OPPOSITE}}$

FIGURE 8-1
Right triangle

$$\text{TANGENT} = \dfrac{\text{SIDE OPPOSITE}}{\text{SIDE ADJACENT}}$$

The other formulas are the inverses of these three. In the machine shop these three will cover most situations.

Figure 8-2 will help to demonstrate the value of triangles in shop mathematics. If the part in Figure 8-2 is to be drilled without using a rotary table, it will be necessary to specify coordinates as dimensioned in Figure 8-2. These are known as *jig borer coordinates*, because they are a common way of locating hole patterns on jig borers. They are also commonly used with milling machines. They are especially important in CNC programming.

The immediate problem in looking at Figure 8-2 is that only the dimensions for holes #1 and #4 are known (they are located on the radius of the bolt circle). However, dimensions a and b can be determined by using trigonometry.

Note that a triangle has been constructed in the first quadrant of the part. If this triangle is solved for the length of its sides, it will supply the information

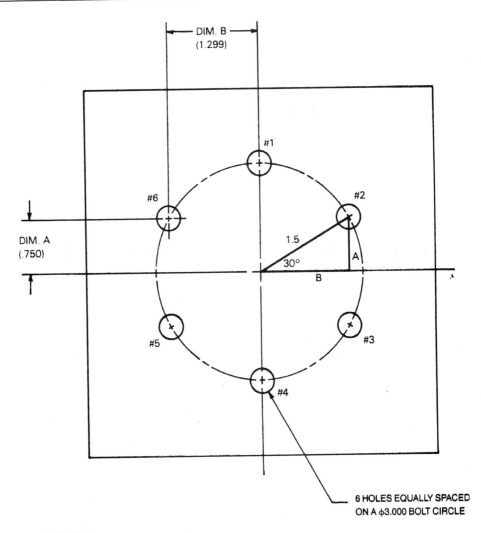

FIGURE 8-2

needed for the missing coordinates. The sides of this triangle are labeled a, b, and 1.5 (which is the hypotenuse of the triangle).

What is known about this triangle? The length of the hypotenuse is half the diameter of the bolt circle, or 1.500 inches. Angle A is also known; since the bolt circle (of 360 degrees) is divided into six equal spaces, the angle between each hole is 60 degrees. Half the distance to each hole lies on each side of the centerline. Therefore, the angle from the centerline to either hole #2 or hole #3 is 30 degrees, which is angle A. The formula for the sine is found to be most practical.

$$\text{SINE A} = \frac{\text{OPPOSITE SIDE}}{\text{HYPOTENUSE}}$$

$$\text{SINE 30} = \frac{a}{1.500}$$

Looking up the sine of 30 degrees in a trigonometric table or using a calculator:

$$.500 = \frac{a}{1.500}$$

$$.500 \times 1.500 = a$$

$$.750 = a$$

To solve for side b, the formula for the cosine is used:

$$\text{COSINE A} = \frac{\text{ADJACENT SIDE}}{\text{HYPOTENUSE}}$$

$$\text{COS A} = \frac{b}{1.500}$$

Using a calculator or table:

$$.866 = \frac{b}{1.500}$$

$$.866 \times 1.500 = b$$

$$1.299 = b$$

The part coordinates are now complete.

Another application of trigonometry is presented in Figure 8-3, where the value of the X dimension is needed. By constructing a triangle as shown in the figure, the length of side b can be determined. If this length is subtracted from the overall length of 3.000 inches, the X dimension is obtained. Two things are known about this triangle: the length of side a and angle A. By using the formula for the tangent, the triangle can be solved as follows:

$$\text{TAN 40} = \frac{\text{OPPOSITE SIDE}}{\text{ADJACENT SIDE}}$$

$$\text{TAN 40} = \frac{1.000}{b}$$

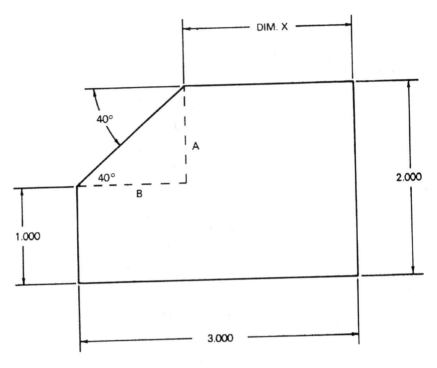

FIGURE 8-3

$$.839 = \frac{1.000}{b}$$

$$.839 \times b = 1.000$$

$$b = \frac{1.000}{.839}$$

$$b = 1.191$$

Dimension X equals 3.000 − 1.191, or 1.809.

USING TRIGONOMETRY FOR CUTTER OFFSETS

A common use for trigonometry in NC programming is calculating cutter offsets for use with linear or circular interpolation (discussed further in Chapter 9). Assume the angle in Figure 8-3 is to be milled. The coordinates of the cutter will have to be determined mathematically because, as Figure 8-4 illustrates, the cutter cannot be positioned at point Y but must be positioned

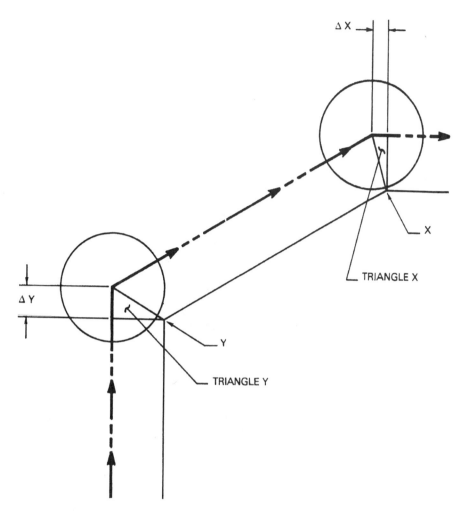

FIGURE 8-4

some unknown distance away. Similarly, the cutter cannot be moved to point X, but some unknown distance short of point X. By solving triangles Y and X, the proper coordinates can be determined.

The angles shown in Figure 8-5 can be found with little effort by looking at the angles formed around points Y and X. When determining angles by this method, three rules must be remembered:

1. The total number of degrees in a circle is 360.
2. The sum of the angles in a triangle is 180 degrees.
3. The complement of an angle is 90 minus the angle.

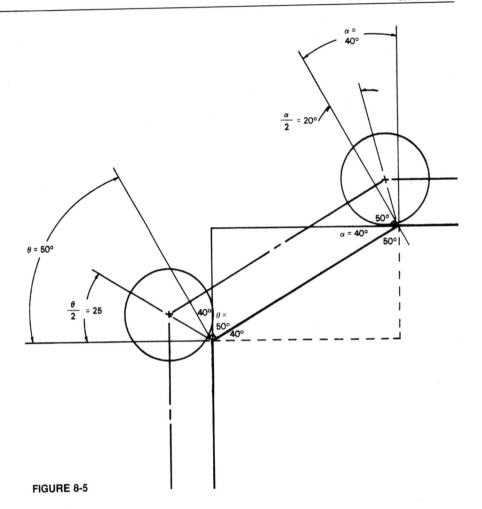

FIGURE 8-5

The angles that will be used for the calculation are 25 degrees for triangle Y and 20 degrees for triangle X, as shown in the figure.

Solving triangle Y for ΔY:

$$\frac{\Delta Y}{.250} = \text{TAN } 25$$

$$\Delta Y = \text{TAN } 25 \, (.250)$$

$$\Delta Y = .46631 \, (.250)$$

$$\Delta Y = .11658 \text{ or } .117$$

Solving triangle X for ΔX:

$$\frac{\Delta X}{.250} = \text{TAN } 20$$

$$\Delta X = \text{TAN } 20 \ (.250)$$

$$\Delta X = .36397 \ (.250)$$

$$\Delta X = .09099 \text{ or } .091$$

The amount of offset can be added to or subtracted from points Y and X to arrive at the correct cutter coordinates. In the next chapter this sort of calculation will be performed for use with linear interpolation.

Refer to Figure 8-6 for right triangle solutions and to Figure 8-7 for oblique-angled triangle solutions.

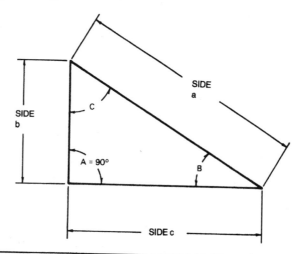

KNOWN VARIABLES	SOLUTION FORMULAS		
SIDE a, ANGLE B	$b = a \times \text{SIN B}$	$c = a \times \text{COS B}$	$C = 90° - B$
SIDE a, ANGLE C	$b = a \times \text{COS C}$	$c = a \times \text{SIN C}$	$B = 90° - C$
SIDE b, ANGLE B	$a = \dfrac{b}{\text{SIN B}}$	$c = b \times \text{COT B}$	$C = 90° - B$
SIDE b, ANGLE C	$a = \dfrac{b}{\text{COS C}}$	$c = b \times \text{TAN C}$	$B = 90° - C$
SIDE c, ANGLE B	$a = \dfrac{c}{\text{COS B}}$	$b = c \times \text{TAN B}$	$C = 90° - B$
SIDE c, ANGLE C	$a = \dfrac{c}{\text{SIN C}}$	$b = c \times \text{COT C}$	$B = 90° - C$
SIDES a AND b	$c = \sqrt{a^2 - b^2}$	$\text{SIN B} = \dfrac{b}{a}$	$C = 90° - B$
SIDES a AND c	$b = \sqrt{a^2 - c^2}$	$\text{SIN C} = \dfrac{c}{a}$	$B = 90° - C$
SIDES b AND c	$a = \sqrt{b^2 + c^2}$	$\text{TAN B} = \dfrac{b}{c}$	$C = 90° - B$

FIGURE 8-6
Solutions of right triangles

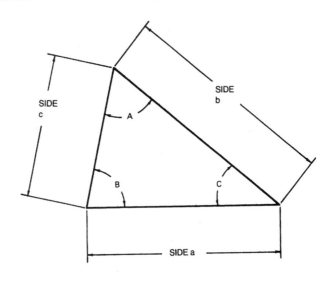

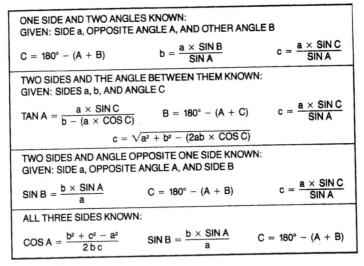

FIGURE 8-7
Solutions for oblique-angled triangles

A MILLING EXAMPLE

Figure 8-8 shows a goblet-shaped casting. A ledge .250 inch deep by 1.000 radius is to be milled in three places, blending to the .120 thick cast web with a .125 radius. A ¼ (.250 diameter) end mill can be used to mill the ledge. Six cutter locations must be calculated. Since three of these locations are mirrors of the other three, only the three locations P1, P2, and P3 need be calculated.

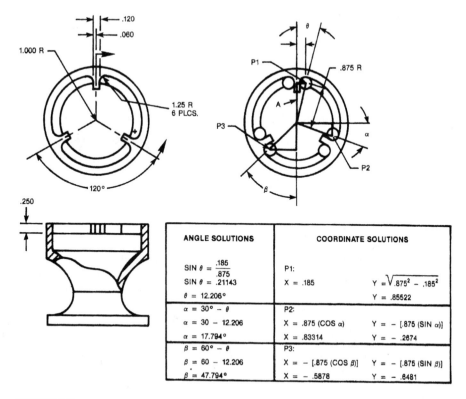

ANGLE SOLUTIONS	COORDINATE SOLUTIONS	
$SIN\ \theta = \dfrac{.185}{.875}$ $SIN\ \theta = .21143$ $\theta = 12.206°$	P1: X = .185	$Y = \sqrt{.875^2 - .185^2}$ Y = .85522
$\alpha = 30° - \theta$ $\alpha = 30 - 12.206$ $\alpha = 17.794°$	P2: X = .875 (COS α) X = .83314	$Y = - [.875\ (SIN\ \alpha)]$ Y = - .2674
$\beta = 60° - \theta$ $\beta = 60 - 12.206$ $\beta = 47.794°$	P3: X = - [.875 (COS β)] X = - .5878	$Y = - [.875\ (SIN\ \beta)]$ Y = - .6481

FIGURE 8-8

There are three triangles that must be solved to determine the proper coordinates. There are two pieces of information known that will allow the solution of P1, and determine angle θ: The radius from the center of the part to the center of the .250 diameter cutter (1.0 R – .125 R) and the .185 (.06 + .125) leg of triangle A. Angle θ can also be determined from this same information as shown in the figure. To determine angle θ, the SIN function is used:

$$SIN\ \theta = \frac{.185}{.875} \text{ or } .21143$$

$$\theta = 12.206 \text{ degrees}$$

Once angle θ is known, angle α and angle β can be determined:

$$\alpha = 30° - \theta \quad \text{and} \quad \beta = 60° - \theta$$

The Pythagorean formula is used to determine the Y coordinate of P1:

$$X = .185 \text{ (known information)}$$

$$Y = \sqrt{.875^2 - .185^2} \text{ or } .85522$$

The coordinates of P2 are calculated by the SIN and COS of angle α.

P2:

$X = .875(COSα)$ OR $.83314$

$Y = - [.875(SINα)]$ or $- .2674$

The coordinates of P3 are calculated by the SIN and COS of angle β.

$X = - [.875(COS β)]$ or $- .5878$

$Y = - [.875(SINβ)]$ or $- .6481$

A LATHE EXAMPLE

A typical lathe programming situation is depicted in Figures 8-9 and 8-10. A 37 degree angle is to be turned, intersecting with a 1.500 diameter. At the intersection points of the angle and part face and the intersection of the angle and the 1.500 diameter, a .005-inch radius is to be generated. This small radius serves to deburr the part. A .015 radius turning tool will be used to turn the part. Most programmers will deburr a part in this manner even if not specifically called for on the blueprint. Most companies have some type of engineering standard that applies in situations such as this and will determine the maximum edge break that is allowed.

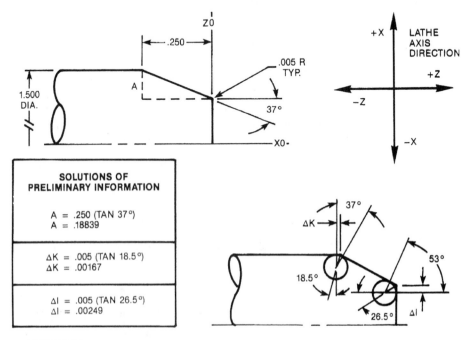

SOLUTIONS OF
PRELIMINARY INFORMATION

A = .250 (TAN 37°)
A = .18839

ΔK = .005 (TAN 18.5°)
ΔK = .00167

ΔI = .005 (TAN 26.5°)
ΔI = .00249

FIGURE 8-9

There are four cutter locations that must be calculated as shown in Figure 8-10: P1, P2, P3, and P4. To determine the cutter locations it is necessary to use a twofold approach. First, A, I, and K must be determined. Second, the coordinates of the four cutter locations are determined using A, ΔI, and ΔK.

To determine the preliminary information:

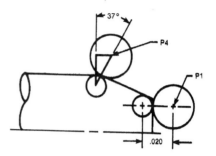

COORDINATE SOLUTIONS

P1:
X = .750 - A - ΔI   Z = -.005 + .020
X = .55912   Z = .015

P2:
X x .55912 + .020 (COS 37°)   Z = -.005 + .02 (SIN 37°)
X = .55912 + .01597   Z = -.005 + .01204
X = .57509   Z = .00704

P3:
X = .745 + .020   Z = -(.250 + ΔK)
X = .765   Z = -.25167

P4:
X = .745 + .020 (COS 37°)   Z = -[.25167 - .020 (SIN 37°)]
X = .745 + .01597   Z = (.25167 - .01204)
X = .76097   Z = -.23963

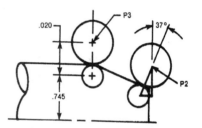

FIGURE 8-10

$$A = .250(TAN\ 37)\ or\ .18839$$

$$\Delta K = .005(TAN\ 18.5)\ or\ .00167$$

$$\Delta I = .005(TAN\ 26.5)\ or\ .00249$$

To determine the cutter locations:

P1:

$$X = .750 - A - \Delta I \qquad Z = -.005 + .020$$

$$X = .55912 \qquad\qquad Z = .015$$

P2:

Note: .55912 in X formula from P1 X calculation.

$$X = .55912 + .020(COS\ 37°) \qquad Z = -.005 + .02(SIN\ 37°)$$

$$X = .57509 \qquad\qquad\qquad Z = .00704$$

P3:

X = .745 + .020   Z = − (.250 + ΔK)

X = .765   Z = −.25167

P4:

Note: .25167 in Z formula from P3 Z calculation.

X = .745 + .020(COS 37°)   Z = −[.25167 − .020(SIN 37°)]

X = .76097   Z = −.23963

It is often necessary when determining coordinates to systematically solve a series of triangles in order to derive the necessary values for the final cutter solution, as this case demonstrates.

SUMMARY

The important concepts presented in this chapter are:

- Right-angle trigonometry is the mathematical science of solving right triangles.
- The sine of an angle equals the side opposite the angle divided by the hypotenuse of the triangle.
- The cosine of an angle equals the side adjacent to the angle divided by the hypotenuse of the triangle.
- The tangent of an angle equals the side opposite the angle divided by the side adjacent to the angle.
- The use of trigonometry is necessary for determining cutter offsets for linear and circular interpolation and for determining other part information from a blueprint.

VOCABULARY INTRODUCED IN THIS CHAPTER

Cosine
Cutter offsets
Jig borer coordinates
Sine
Tangent
Trigonometry

REVIEW QUESTIONS

1. What is the sine of an angle? The cosine? The tangent?
2. What are the sine, cosine, and tangent of the triangle in Figure 8-11? What is angle B?

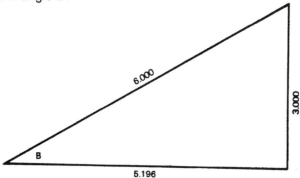

FIGURE 8-11
Triangle for review question #2

3. What are the coordinates of the holes in the part in Figure 8-12, assuming that the center is X0/Y0?

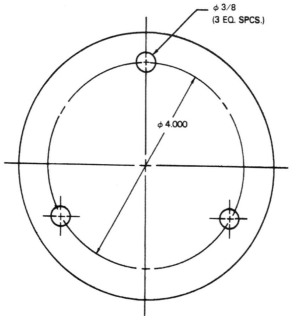

FIGURE 8-12
Part drawing for review question #3

4. What are the coordinates of the four cutter locations indicated in Figure 8-13?

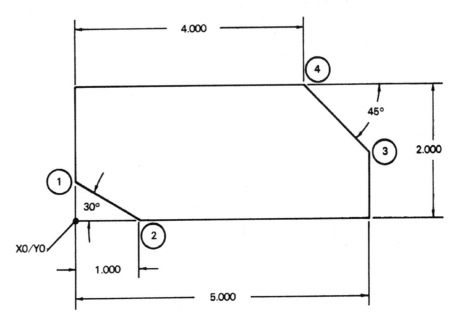

FIGURE 8-13
Part drawing for review question #4

CHAPTER 9

Linear and Circular Interpolation

OBJECTIVES: Upon completion of this chapter, you will be able to:

- Write programs using linear interpolation to cut simple angles.
- Write simple programs using circular interpolation to mill arcs.

LINEAR INTERPOLATION

Linear interpolation simply means cutting a straight line between two points. Sometimes this is referred to as a feedrate move, since modern CNC controls automatically perform linear interpolation on any move made while in feedrate mode. Prior to the advent of modern CNC controls, special codes were necessary to turn on the built-in linear interpolation system. Some CNC controls also will interpolate rapid moves, while others simply move the axes drive motors at maximum speed in rapid traverse mode.

Machines capable of linear interpolation have a continuous-path control system, meaning that the drive motors on the various axes can operate at varying rates of speed. Virtually all modern CNC controls utilize continuous path controls. When cutting an angle, the MCU calculates the angle based on the programmed coordinates. Since the MCU knows the current spindle location, it can calculate the difference in the X coordinate between the current position and the programmed location. The change in the Y coordinate divided by the change in the X coordinate yields the slope of the cutter centerline path.

Calculating Cutter Offsets

Figure 9-1 shows a part on which an angle is to be milled. Figure 9-2 shows the cutter path and geometric relationships between the cutter and the part. The cutter has already been positioned at location #1, Figure 9-2. A .500-inch diameter end mill is being used. Before the angle can be cut, it is necessary to first position the spindle at location #2, Figure 9-2. Notice that the Y-axis coordinate for location #2, as dimensioned on the part, is not

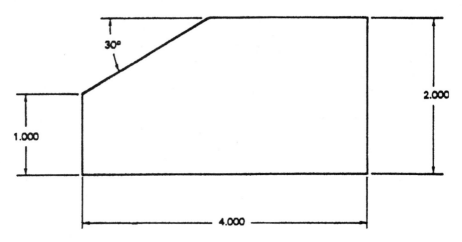

FIGURE 9-1
Part drawing

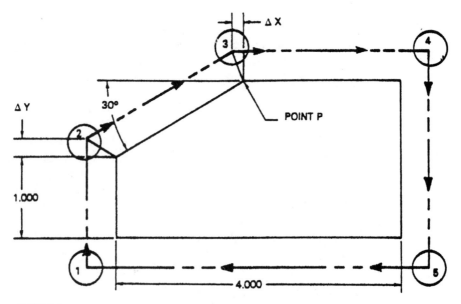

FIGURE 9-2
Cutter path for part in Figure 9-1

the same point as the edge of the angle. In order to determine this Y-axis cutter offset, it will be necessary to determine the amount that must be added to the dimension on the part print to place the spindle at location #2. Similarly, it will be necessary to calculate an amount to be subtracted from the point on the part designated as "P" to arrive at the X-axis coordinate for location #3. This is the same cutter offset situation presented in Chapter 8. Figure 9-3 represents an enlarged view of locations #2 and #3, illustrating the triangles involved in determining the offsets.

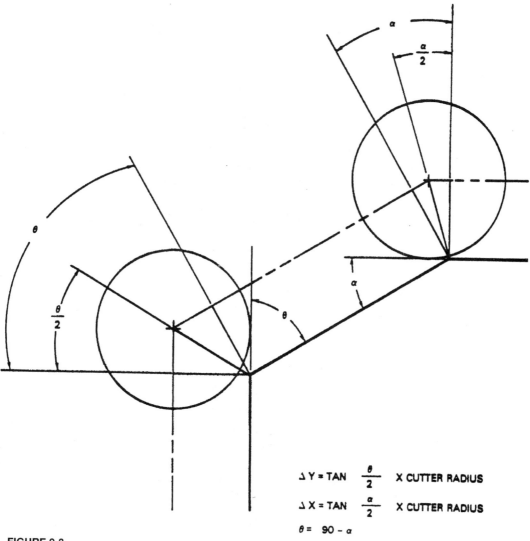

$$\Delta Y = TAN \ \frac{\theta}{2} \ \ X \ CUTTER \ RADIUS$$

$$\Delta X = TAN \ \frac{\alpha}{2} \ \ X \ CUTTER \ RADIUS$$

$$\theta = \ 90 - \alpha$$

FIGURE 9-3
Determining cutter offset

Calculating ΔY

The formulas from Appendix 6, Figure 1, can be used to determine the offsets as follows:

$$\Delta Y = CR[TAN(\alpha/2)] \quad \text{where CR = Cutter radius}$$
$$\Delta Y = .25[TAN(30)]$$
$$\Delta Y = .25(.5774)$$
$$\Delta Y = .144$$

The ΔY offset to be added to the part dimension to arrive at the Y coordinate for location #2 is .144.

Calculating ΔX

The offset for location #3 can be determined as follows:

$$\Delta X = CR[TAN(\alpha/2)]$$
$$\Delta X = .25[TAN(15)]$$
$$\Delta X = .25(.26794)$$
$$\Delta X = .067$$

Before using this information to determine the X-axis coordinate, it also will be necessary to calculate the coordinate location of point "P" along the X axis. Again, using the trigonometry formulas, the coordinate can be calculated as follows:

$$TAN(30) = 1.000/b$$
$$.5774 = 1.000/b$$
$$.5774 \times b = 1.000$$
$$b = 1.000/.5774$$
$$b = 1.732$$

Subtracting .067 (the ΔX offset) from 1.732 produces the X-axis coordinate for the cutter of 1.665. The ΔY offset, which was found earlier to be .144 can now be added to the 1.000 Y-axis dimension on the part to arrive at a Y-axis coordinate of 1.144.

Interpolating in Word Address Format

Milling an angle with word address is not complicated. Since the interpolator is automatically turned on when feedrate mode is commanded, milling simply becomes a matter of specifying the coordinates, along with the G01 feedrate mode code. In CNC shops, G01 is sometimes called the *feedrate mode code*. Other times it is called the *linear interpolation code*. With modern CNC controls, the terms mean the same thing. Any feedrate move is considered an interpolated angled line move. A move along the X axis only would cut an angled line of 0 degrees. A move along the Y axis would cut an angle of 90 degrees.

To illustrate linear interpolation, the following program lines would move the cutter from location #1 to locations #2, #3, and #4.

```
N...G01Y1.144        (move from #1 to #2)
N...X1.665Y2.25      (move from #2 to #3)
N...X4.25            (move from #3 to #4)
```

A G01 is given to turn on linear interpolation (feedrate mode). The coordinate Y1.144 moves the cutter to location #2. The X1.665, Y2.25 coordinates calculated earlier are programmed to move the cutter from location #2 to #3. The X4.25 coordinates then move the cutter from #3 to #4. Remember, G01 is a modal code. The machine remains in linear interpolation mode (feedrate mode) for all the coordinates specified. The G01 is active until cancelled by another motion mode G code (G00, G02, G03, or G04).

Additional Example

Linear interpolation is not difficult. Aside from calculating the cutter offsets necessary to position the spindle, it is the same as straight line milling. The only real difference is that an X and a Y coordinate are specified for the ending point of the angle since there is a change in position of both axes.

Figure 9-4 shows another cutter offset situation. This part has two angles which intersect each other. In this case, the calculation of the cutter offsets becomes somewhat more complicated. In order to program locations #1 and #3, the formula in Appendix 6, Figure 1, can be used. For location #2, the formula from Appendix 6, Figure 2, can be used. A .500-inch cutter will be used.

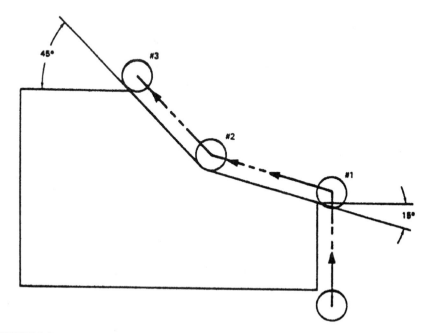

FIGURE 9-4
Part drawing with cutter path shown

Offset For Location #1

$$\Delta X = CR$$

$$\Delta Y = CR[TAN(75/2)]$$
$$\Delta Y = .25[TAN(37.5)]$$
$$\Delta Y = .25(.7673)$$
$$\Delta Y = .1918$$

Offset for Location #2

$$\Delta X = CR \times \{ [SIN((45+15)/2)] / [COS((45-15)/2)] \}$$
$$\Delta X = .25 \times [SIN(30)/COS(15)]$$
$$\Delta X = .25 \times (.5/.9659)$$
$$\Delta X = .25 \times .5176$$
$$\Delta X = .1294$$
$$\Delta Y = CR \times \{ [COS((45+15)/2] / [COS((45-15)/2)] \}$$
$$\Delta Y = .25 \times [COS(30)/COS(15)]$$
$$\Delta Y = .25 \times (.866/.9659)$$
$$\Delta Y = .25 \times .8966$$
$$\Delta Y = .2241$$

Offset for Location #3

$$\Delta X = CR \times [TAN(45/2)]$$
$$\Delta X = .25 \times TAN(22.5)$$
$$\Delta X = .25 \times .4142$$
$$\Delta X = .1036$$
$$\Delta Y = CR$$

Other cutter situations will present themselves in CNC part programming, such as arcs tangent to an angle, or arcs tangent to other arcs. A good working knowledge of the formulas listed in Figures 8-6 and 8-7 and the figures in Appendix 6 should be developed by the prospective CNC part programmer. CAM programming systems automatically calculate cutter offsets with speed and accuracy no programmer can match. For this reason, CAM systems have become the preferred programming system in many shops. A good programmer or CNC operator must still know how to calculate cutter offsets in order to edit programs in the machine control during the first piece setup.

CIRCULAR INTERPOLATION

In cutting arcs, the MCU uses its ability to generate angles to approximate an arc. Since the machine axes do not revolve around a centerpoint in a typical three-axis arrangement, the cutting of a true arc is not possible. Circular interpolation is the term used to describe generating a move consisting of a series of straight-line chord segments by the MCU in two axes to simulate circular motion, as illustrated in Figure 9-5.

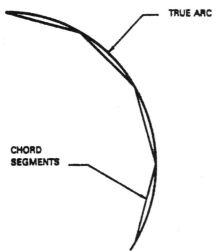

FIGURE 9-5
Circular interpolation chord segments

These chord segments are very small and practically indistinguishable from a true arc.

Figure 9-6 shows a part with a radius to be machined. Figure 9-7 shows the cutter path. In order to generate the radius, circular interpolation will be used to send the cutter from location #3 to location #4, Figure 9-7. A .500-inch diameter end mill will be used.

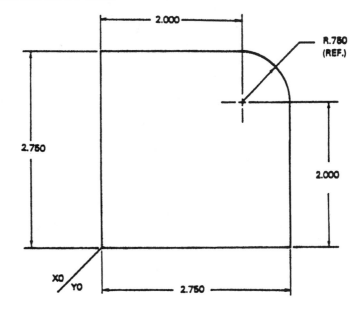

FIGURE 9-6
Part with radius to be machined

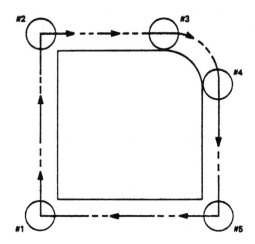

FIGURE 9-7
Cutter path for part shown in Figure 9-6

To generate an arc, the MCU needs to know the following information.

1. The axes to be used in generating the arc.
2. The direction of interpolation, clockwise or counterclockwise.
3. The starting X/Y/Z coordinate of the arc.
4. The ending X/Y/Z coordinates of the arc.
5. The X/Y/Z coordinates of the arc centerpoint.

The formats for circular interpolation in word address format follows.

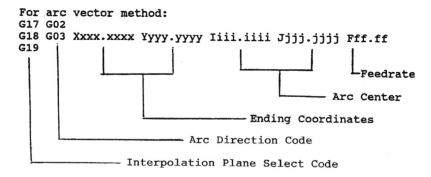

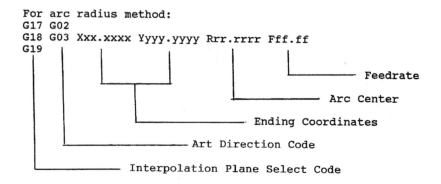

An explanation of these formats is contained in the discussion that follows.

Specifying Axis for Interpolation

Circular interpolation by definition involves only two axes. On FANUC-style controls, a plane designation code is used to select which pair of axis will be used to generate the arc motion. There are three G codes used to specify these planes:

- G17 – Selects the X/Y plane (X and Y axis)
- G18 – Selects the Y/Z plane (Y and Z axis)
- G19 – Selects the Z/X plane (Z and X axis)

These G codes are modal. A G17, for example, is cancelled only by a G18 or G19.

The X/Y plane (using the X and Y axis) is the most common orientation for circular interpolation, therefore, G17 will be used throughout the examples in this text.

Specifying Arc Direction

Circular interpolation can be accomplished in one of two directions: clockwise, or counterclockwise. There are two G codes used to specify direction.

- G02 - circular interpolation clockwise (CLW)
- G03 - circular interpolation counterclockwise (CCLW)

G02/G03 codes are modal. They will cancel an active G00 (rapid traverse) or G01 (linear interpolation) codes. G02/G03 are feedrate mode codes, just as G01 is. The difference lies in the type of interpolation used. G01 generates straight-line interpolation motion. G02/G03 generates arc simulation interpolation motion.

Specifying Beginning and Ending Arc Coordinates

The MCU requires the spindle be positioned at the start of the arc when the G02/G03 command is given. The current spindle position is the beginning arc coordinates. The axis coordinates given on the G02/G03 line are the spindle ending points of the arc motion.

Specifying Arc Centerpoints

There are two methods used to specify arc centerpoints: arc vector method and radius method. The arc vector method involves specifying the coordinates of the arc centerpoint as X/Y values. In the radius method, the arc centerpoint is calculated internally by the MCU. The programmer simply specifies the radius value required.

Arc Vector Method

Since X, Y, and Z addresses are used to specify the end point of an arc, secondary addresses are required to specify the centerpoint of an arc. The following addresses are used to designate arc centerpoints.

I – X axis coordinate of an arc.
J – Y axis coordinate of an arc.
K – Z axis coordinate of an arc.

Since circular interpolation occurs only in two axes, only two of these three codes will be required to generate an arc. When using the X/Y plane for milling arcs, as this text does, the I and J addresses are used.

The different ways controllers required the arc centerpoints to be specified complicate this matter: *absolute coordinates, to circle center,* or *from circle center.* FANUC-style controls usually utilize the *to circle center* method.

Absolute Coordinates

Some controls require the arc centerpoints specified by I, J, and/or K be the position of the arc center relative to the coordinate system origin. In other words, the center of the arc is specified just as if it were a cutter coordinate using absolute positioning. In Figure 9-6, the arc centerpoints are at X2.000, Y2.000. They would be specified as I2.0000 J2.0000 as in the following circular interpolation block.

N120 G17 G02 X3. Y2. I2. J2. F7.

To Circle Center

Some controls require the arc center points be specified as an incremental coordinate, looking from the center of the cutter to the center of the circle. In Figure 9-6, the radius of the arc is .750. The radius of .500 diameter end mill is .25 inch. To specify the centerpoint of the arc when the cutter is positioned at location #3, Figure 9-7 the incremental value of 0.0000 inch in X and −1.000 inch in Y would be specified as I0.0000 J-1.0000 as given in the following block of CNC code.

N120 G17 G02 X3. Y2. I0. J-1. F7.2

The 1.000 incremental J value is calculated by adding the .250 inch cutter radius to the .750 part radius. A minus value is required since the direction from the cutter centerline to the arc centerline is in a minus direction. The spindle is really generating a 1.000-inch arc when the cutter center is taken into account.

From Circle Center

The *from circle center* method is the same as the *to circle center* except the incremental coordinate is specified looking from the center of the arc to the center of the cutter. The signs associated with the I, J, and K addresses will be the reverse of the *to circle center* method. The following line of code specifies the arc coordinates when the cutter is positioned at location #3, Figure 9-7.

N120 G17 G02 X3. Y2. I0. J1. F7.2

Notice that the only difference between this block of code and the one given previously is the sign of the J address.

Radius Method

When using the radius method, the programmer only needs to specify the radius to be cut when programming the cut. Instead of using the I, J, and/or K addresses, the R address is used to specify the arc radius. The following block of CNC code moves the cutter from location #3, Figure 9-7 to location #4 using the radius method.

N120 G17 G02 X3. Y2. R1. F7.2

Notice that the radius to be cut is still 1.000 inch. The controller is commanding motion of the spindle centerline. It does not know that there is a .500-inch diameter cutter in the spindle. The true cutter path is still a 1.000-inch arc.

Although the radius method is easier to use than the arc vector method, the latter method is still common. This is most likely because the radius method became available only with the advent of modern CNC controllers. Many of today's programming practices have ties to the tape-controlled MCU of days gone by. This use of the arc vector method is one of these.

Milling the Arc

Putting together all these pieces, the following sections of CNC code will mill the part surface in Figures 9-6 and 9-7.

Arc Vector Method

The following code uses absolute coordinates to specify arc centerpoints.

```
N001 G80 G90 G00 G98              (safety block)
N100 T01 M06                      (tool change block)
N101 G00 X-.25 Y-.25              (rapid to location #1)
N102 G44 Z-1. H01                 (tool offset pickup)
N103 G01 Y3.                      (feed #1 to #2)
N104 X2.                          (feed #2 to #3)
N105 G17 G02 X3. Y2. I2. J2. F7.2 (circular move to #4)
N106 G01 Y-.25                    (feed #4 to #5)
N107 X-.25                        (feed #5 to #1)
N108 G00Z0                        (retract Z)
N109 G49                          (cancel tool offset)
N110 G91 G28 X0. Y0. Z0.          (rapid to home zero)
N110 M01                          (optional stop code)
```

The following code uses the *to circle center* method to specify arc centerpoints. This example is typical of FANUC, General Numeric, and other FANUC-style controls.

```
N001 G80 G90 G00 G98          (safety block)
N100 T01 M06                  (tool change block)
N101 G00 X-.25 Y-.25          (rapid to #1)
N102 G44 Z-1. H01             (tool offset pickup)
N103 G01 Y3.                  (feed #1 to #2)
N104 X2.                      (feed #2 to #3)
N105 G17 G02 X3. Y2. I2. J-1. F7.2   (circular move to #4)
N106 G01 Y-.25                (feed #4 to #5)
N107 X-.25                    (feed #5 to #1)
N108 G00Z0                    (retract Z)
N109 G49                      (cancel tool offset)
N110 G91 G28 X0. Y0. Z0.      (rapid to home)
N110 M01                      (optional stop code)
```

The following code uses the *from circle center* method to specify arc centerpoints.

```
N001 G80 G90 G00 G98          (safety block)
N100 T01 M06                  (tool change block)
N101 G00 X-.25 Y-.25          (rapid to #1)
N102 G44 Z-1. H01             (tool offset pickup)
N103 G01 Y3.                  (feed #1 to #2)
N104 X2.                      (feed #2 to #3)
N105 G17 G02 X3. Y2. I2. J-1. F7.2   (circular move to #4)
N106 G01 Y-.25                (feed #4 to #5)
N107 X-.25                    (feed #5 to #1)
N108 G00Z0                    (retract Z)
N109 G49                      (cancel tool offset)
N110 G91 G28 X0. Y0. Z0.      (rapid to home)
N110 M01                      (optional stop code)
```

Radius Method

The following code mills the part using the *radius* method of specifying arc center coordinate.

```
N001 G80 G90 G00 G98          (safety block)
N100 T01 M06                  (tool change block)
N101 G00 X-.25 Y-.25          (rapid to #1)
N102 G44 Z-1. H01             (tool offset pickup)
N103 G01 Y3.                  (feed #1 to #2)
N104 X2.                      (feed #2 to #3)
N105 G17 G02 X3. Y2. R1. F7.2 (circular move to #4)
N106 G01 Y-.25                (feed #4 to #5)
N107 X-.25                    (feed #5 to #1)
N108 G00Z0                    (retract Z)
N109 G49                      (cancel tool offset)
N110 G91 G28 X0. Y0. Z0.      (rapid to home)
N110 M01                      (optional stop code)
```

Additional Circular Interpolation Examples

The programs just discussed deal with simple arcs which intersect a line parallel to a machine axis. In many cases, however, an arc will intersect an angle or another arc. Figures 9-8 and 9-9 are examples of such cases. The cutter offsets for these situations can be found by using the formulas from Appendix 6. The cutter radius (CR) in the following examples is .250 inch.

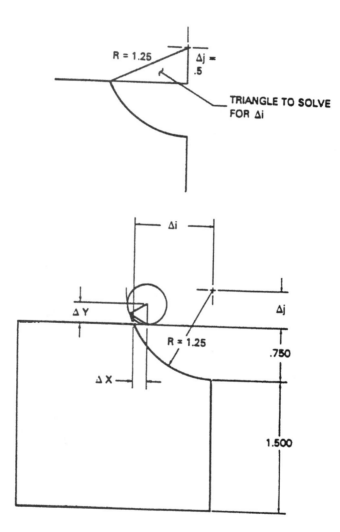

FIGURE 9-8
Part drawing for additional example

To calculate ΔX and ΔY in Figure 9-8, it is necessary to calculate Δi and Δj:

$$\Delta j = 1.25 - .75$$
$$\Delta j = .5$$

Using the Pythagorean theorem, i can be determined:

$$\Delta i = SQRT[(1.25^2 - .5^2)]$$
$$\Delta i = SQRT[(1.562 - .25)]$$
$$\Delta i = SQRT(1.312)$$
$$\Delta i = 1.1454$$

This information can then be used to determine X and Y:

$$\Delta Y = CR$$
$$\Delta X = \Delta i - SQRT[(R - CR)^2 - (\Delta j - CR)^2]$$
$$\Delta X = 1.1454 - SQRT[(1.125 - .25) - (.5 - .25)]$$
$$\Delta X = 1.1454 - SQRT(1 - .0625)$$
$$\Delta X = 1.1454 - SQRT(.9375)$$
$$\Delta X = 1.1454 - .96825$$
$$\Delta X = .17715$$

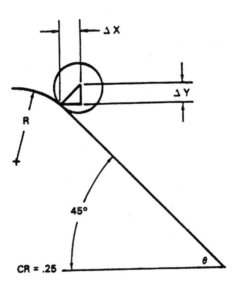

FIGURE 9-9

To calculate ΔX and ΔY in Figure 9-9:

$$\Delta X = CR \times SIN(45)$$
$$\Delta X = .25 \times (.7071)$$
$$\Delta X = .1769$$

$$\Delta Y = CR \times COS(45)$$
$$\Delta Y = .25 \times (.7071)$$
$$\Delta Y = .1769$$

In this case, since the angle is 45 degrees, the offsets for ΔX and ΔY are the same. Had a different angle been used, the offsets would be different.

A COMPREHENSIVE EXAMPLE

Figure 9-10 depicts a switch housing that is to be milled on a three-axis vertical machining center with a FANUC 6M control. The accompanying program serves as an example of a program using both linear and circular interpolation. There are no straight lines on this part. All contouring moves will involve angles or arcs.

The setup for this part involves two fixtures as shown on the setup sheet in Figure 9-11. The first fixture involves holding a slug of round stock in an OD collet. The second fixture holds the part in an OD collet on the 2.000 diameter which has been machined on the first fixture. The summarized sequence of operations follows:

First Fixture

1. Rough and finish the top of the part, using a circular motion with a 1.0 dia. end mill to achieve cutte coverage.
2. Rough and finish the 1.625 dia. ID with 1.0 end mill.
3. Rough and finish the 2.000 dia. OD with 1.0 end mill.
4. Rough drill the .188 dia. holes to .177 dia. with .177 dia. stub drill.
5. Finish ream the .188 dia. holes to .188 dia. with a .188 dia. reamer.

Second Fixture

1. Rough and finish top of part to establish 1.375 overall dimension with a 1.0 dia. end mill.
2. Rough and finish the outside periphery of the part using a .625 dia. end mill.

The program manuscript is given in Figure 9-12. The circular interpolation moves in this program use the cutter vector method of specifying arc center coordinates. The arc centers are specified using *to circle center* coordinates, as is commonplace with FANUC-style controls.

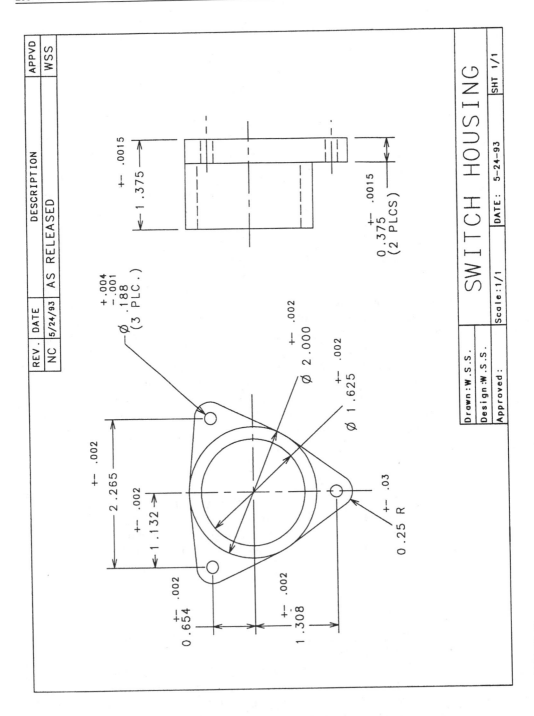

FIGURE 9-10
Part drawing for comprehensive example

STA. NO.	TOOL DESCRIPTION
1.	1" END MILL, 4-FLT CENTER CUTTING (Ø1.000)
2.	Ø.177 STUB DRILL
3.	Ø.1880 REAMER
4.	5/8 END MILL, 2FLT (Ø.625)

REMARKS

TAPE NUMBER: 9012
FIXTURE: C-1232 & C-1233 COLLETS
WORK COORDINATES/FIXTURE OFFSETS

92 FIX 1	92 FIX 2
C/L COLLET	C/L COLLET
C/L COLLET	C/L COLLET
2" ABOVE BANK	2" ABOVE BANK

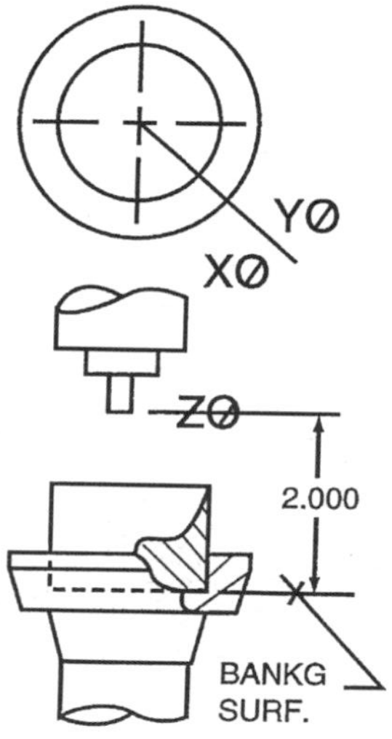

FIX #1 USES OFFSETS H01, H02, H03

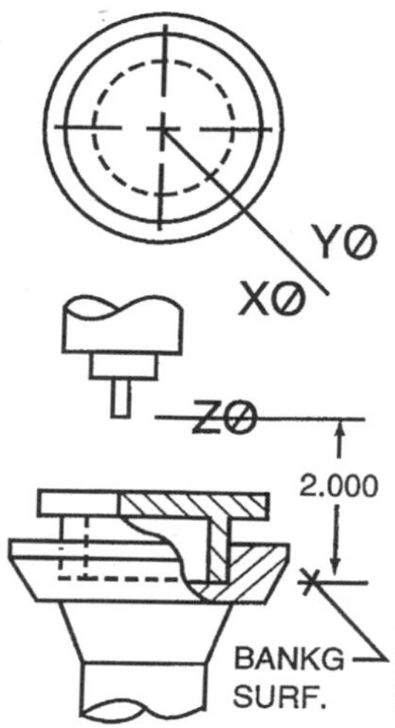

FIX #2 USES OFFSETS H12 & H04

2.000
BANKG SURF.
X0   Y0   Z0

DRWN: WSS
PRGM: WSS
APVD: 7/20/93

PART NUMBER: SWITCH HOUSING   OP: 20
MACHINE: UNIVERSAL VNC

FIGURE 9-11
Setup sheet for part in Figure 9-10

```
%
O9012
(SWITCH HOUSING OPERATION 30 06/18/93 07:11:54)
N1G91G28Z.0G28X.0Y.0
N2G28X.0Y.0Z.0
N3G92X-12.752Y-7.453Z.0
(* *********)
(* TOOL NO. 1)
(* 1.000 DIA. END MILL)
(* ROUGH/FINISH TOP OF PART)
(* *********)
N4T01M06
N5G90G00G80
N6G00X.0Y.0S1800M03
N7G44Z-.5H01M08
(ROUGH THE TOP OF THE PART)
(LEAVE .010 STK. TO FINISH)
N8G01Z-1.615F8.
N9Y.5F20.
N10G17G03X.0Y.5I.0J-.5
N11G01Y.0
(FINISH TOP OF PART)
N12S3000M03
N13Z-1.625F16.
N14Y.5
N15G03X.0Y.5I.0J-.5
N16G01Y.0
(ROUGH 1.625 DIA. BORE)
N17S1800M03
N18Z-.615F7.
N19Y.3025
N20G03X.0Y.3025I.0J-.3025
N21G01Y.2025
(FINISH 1.625 DIA. BORE)
N22Z-.625F15.
N23Y.3125
N24G03X.0Y.3125I.0J-.3125
N25G01Y.0
N26G00Z.0
(ROUGH 2.000 DIA.)
N27Y-3.
N28Z-.615
N29G01Y-1.51
N30G02X.0Y-1.51I.0J1.51
N31G01Y-1.61
(FINISH 2.000 DIA.)
N32Z-.625
N33Y-1.5
N34G02X.0Y-1.5I.0J1.5
N35G01Y-3.
N36G49M09
N37G91G00G28Z.0
N38G28X.0Y.0
N39M01
(* *********)
(* TOOL NO. 2)
(* .177 DIA. STUB DRILL)
```

FIGURE 9-12 *(Continues to page 210)*

```
(* DRILL HOLES IN FLANGE)
(* *********)
N40T02M06
N41G90G00G80
N42G00X1.132Y.654S2200M03
N43G44Z-.525H02M08
N44G81G98X1.132Y.654Z-2.125R-1.525F8.8
N45X-1.132
N46X.0Y-1.308
N47G80
N48G49M09
N49G91G00G28Z.0
N50G28X.0Y.0
N51M01
(* **********)
(* TOOL NO. 3)
(* .188 DIA. REAMER)
(* FINISH REAM HOLE PATTERN)
(* **********)
N52T03M06
N53G90G00G80
N54G00X1.132Y.654S1200M03
N55G44Z-.525H03M08
N56G85G98X1.132Y.654Z-2.075R-1.525F9.6
N57X-1.132
N58X.0Y-1.308
N59G80
N60G49M09
N61G91G00G28Z.0
N62G28X.0Y.0
N63M01
(* **********)
(* SET COORDINATE SYSTEM FOR 2ND FIXTURE)
(* **********)
N64G28X.0Y.0Z.0
N65G92X-8.253Y-7.253Z.0
(* **********)
(* RECALL TOOL 1)
(* 1.000 DIA. END MILL)
(* MILL TOP OF PART ON FIXTURE 2)
(* USES LENGTH OFFSET H11)
(* **********)
N66T01M06
N67G90G00G80
N68G00X.0Y3.S2000M03
N69G44Z-.615H11M08
(ROUGH TOP OF PART)
N70G01Y.5F16.
N71G02X.0Y.5I.0J-.5
N72G01Y.0
(FINISH TOP OF PART)
N73Z-.625
N74Y.5
N75G03X.0Y.5I.0J-.5
N76G01Y3.
N77G49M09
N78G91G00G28Z.0
```

```
N79G28X.0Y.0
N80M01
(* **********)
(* TOOL NO. 4)
(* .625 DIA. END MILL)
(* MILL OUTSIDE PERIPHERY OF PART)
(* **********)
N81T04M06
N82G90G00G80
N83G00X.0Y3.S2200M03
N84G44Z-1.15H04M08
(ROUGH OUTSIDE)
N85G01Y1.3225Z-1.1F12.
N86G02X.1151Y1.3175I.0J-1.3225
N87G01X1.1818Y1.2243
N88G02X1.651Y.4123I-.0498J-.5703
N89G01X1.1988Y-.5584
N90G02X1.0835Y-.7583I-1.1988J.5584
N91G01X.469Y-1.6363
N92G02X-.469Y-1.6363I-.469J.3283
N93G01X-1.0835Y-.7583
N94G02X-1.1988Y-.5584I1.0835J.7583
N95G01X-1.651Y.4123
N96G02X-1.1818Y1.2243I.519J.2417
N97G01X-.1151Y1.3175
N98Y1.4175
(FINISH OUTSIDE)
N99G00X-1.6945Y1.2699
N100G01Y1.1695
N101X-.1142Y1.3075
N102G02X.1142Y1.3075I.1142J-1.3075
N103G01X1.1809Y1.2144
N104G02X1.6419Y.4165I-.0489J-.5604
N105G01X1.1898Y-.5542
N106G02X1.0753Y-.7526I-1.1898J.5542
N107G01X.4608Y-1.6305
N108G02X-.4608Y-1.6305I-.4608J.3225
N109G01X-1.0753Y-.7526
N110G02X-1.1898Y-.5542I1.0753J.7526
N111G01X-1.6419Y.4165
N112G02X-1.1809Y1.2144I.5099J.2375
N113G01X.1Y1.3262
N114G00Z.0
N115G49M09
N116G91G00G28Z.0
N117G28X.0Y.0
N118M01
N119G91G00G28X.0Y.0Z.0
N120M30
%
```

FIGURE 9-12
Program to machine part in Figure 9-11

SUMMARY

The important concepts presented in this chapter are:
- Linear interpolation is the ability to cut angles. It is simply a feedrate move, in a straight line, between two points.
- Circular interpolation is the ability to cut arcs or arc segments. Arcs are cut by means of a series of chordal segments generated by the MCU to approximate the arc curvature.
- It is necessary to calculate the cutter offset coordinates when using linear and circular interpolation.
- G01 is the code to institute linear interpolation. It also is referred to as the feedrate move code.
- G02 and G03 are used to institute circular interpolation. G02 turns on clockwise interpolation. G03 turns on counterclockwise interpolation.
- There are two methods used to specify the arc centerpoints to the MCU: the arc vector method and the radius method.
- When using the arc vector method to specify centerpoints, some controls require the centerpoints to be given in absolute coordinates, some in incremental coordinates from the cutter center to the circle center, and others in incremental coordinates from the circle center to the cutter center.
- The format for circular interpolation for the arc vector method is:
    G17 G02
    G18 G03 X... Y... Z... I... J... K...
    G19

For most uses (X/Y plane interpolation) the format is:
    G17 G02/G03 X... Y... I... J...
- The format for circular interpolation for the radius method is:
    G17 G02
    G18 G03 X... Y... Z... R...
    G19

For most uses (X/Y plane interpolation) the format is:
    G17 G02/G03 X... Y... R...

VOCABULARY INTRODUCED IN THIS CHAPTER

Arc centerpoints
Arc vector method
Circular interpolation
From circle center
Linear interpolation
Radius method
To circle center

REVIEW QUESTIONS

1. What will be the result of cutting the angle in Figure 9-13 if the Y offset is not calculated, by the 4.000 dimension is used?
2. What two formulas can be used in calculating coordinates for simple angles, where the angle intersects a line parallel to a machine axis?
3. What code is used to initiate linear interpolation?
4. What is the format for circular interpolation using the arc vector method? The radius method?
5. What are I and J used for? What is R used for?

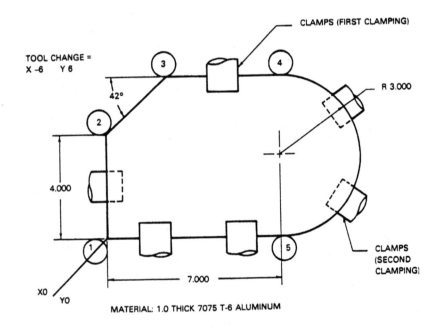

FIGURE 9-13
Part drawing for review question #6

6. Write a program to mill the part in Figure 9-13.

CHAPTER 10

Cutter Diameter Compensation

OBJECTIVES: Upon completion of this chapter, you will be able to:

- Define cutter diameter compensation.
- Describe ramp on and ramp off moves and explain their importance.
- List the precautions necessary when using cutter diameter compensation.
- Write programs in word address and Machinist Shop Language that utilize cutter diameter compensation.

DEFINITIONS AND CODES

Programs presented in previous chapters required an allowance for the cutter radius in the programmed coordinates. Some types of CNC machinery have a built-in feature called cutter diameter *compensation (cutter comp)* that allows the part line to be programmed. (Confusion may be caused by use of the terms "offset" and "compensation." In this text, "compensation" refers to cutter diameter offset. The term "offset" refers to tool length offset and the change in axis coordinates when programming arcs and angles.) Cutter comp is also called cutter radius offset (CRO) by some controller manufacturers. In computer-aided programming languages (such as APT) and some CAD/CAM systems it is also called cutcom. These terms all refer to the same thing: a built-in cycle in the MCU that, when activated, alters the tool path by an amount contained in the cutter comp register. The value in the register is entered in by the setup person when the job is being prepared.

Cutter comp is accomplished through the use of G codes: G40, G41, G42.

G40—Cutter diameter compensation cancel. Upon receiving a G40, cutter diameter compensation is turned off. The tool will change from a compensated position to an uncompensated position on the next X, Y, or Z axis move.

G41—Cutter diameter compensation left. Upon receiving a G41, the tool will compensate to the left of the programmed surface. The tool will move to a compensated position on the next X, Y, or Z axis move after the G41 is received.

213

G42—Cutter diameter compensation right. Compensates to the right of the programmed surface.

Most controllers allow compensation to be performed on any two axes. A G code is used to determine which axes combination is to be used. If the part is to be machined using the X and Y axes, compensation is desired in the X/Y plane. If using the X and Z axes, compensation in the Z/X plane is needed. If using the Y and Z axes, compensation is needed in the Y/Z plane. The X/Y plane is used most commonly. The G codes used to select the desired work plane are:

G17—X/Y plane.
G18—Z/X plane.
G19—Y/Z plane.

Two terms are important for understanding cutter diameter compensation: *ramp on* and *ramp* off. Figure 10-1 will help to illustrate their meaning. If a tool is moved from point #1 to point #2 following a G41 command, the cutter will compensate in a plane perpendicular to the part surface and the spindle will move to point #3 rather than point #2. This initial compensation move is called the *ramp on* move. The machine is in the process of adjusting its path for the entire move from point # 1 to point #2. By the time it reaches point #2, it has fully compensated its path.

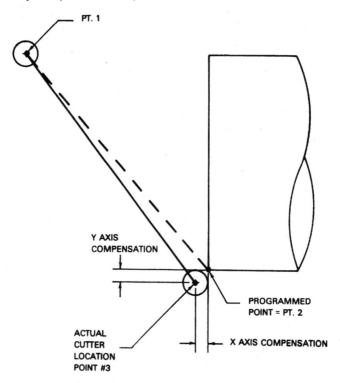

FIGURE 10-1
Ramp on move

In Figure 10-2 another type of situation is demonstrated. The cutter started at point # 1 in this illustration, presenting the possibility that the corner of the part might be cut off in the process of moving to point #2 if the spindle were moving downward or already positioned there. If point #1 was the desired tool change location for this part, the spindle would need to be fully retracted and lowered after reaching the programmed location. When clamps or fixturing devices do not interfere, it is not uncommon to rapid the Z axis to depth on the same move that positions X and Y. Some controllers do not allow cutter comp to be instituted in two axes simultaneously. In these cases it is necessary to program a location away from the part surface, and ramp on the compensation 90 degrees from the desired part surface, as in Figure 10- 3. Controllers are often particular about the manner in which cutter comp is ramped on. It is advisable when programming several different controllers or controllers of different ages to use the method in Figure 10-3. This is the most successful method of ramping on cutter comp.

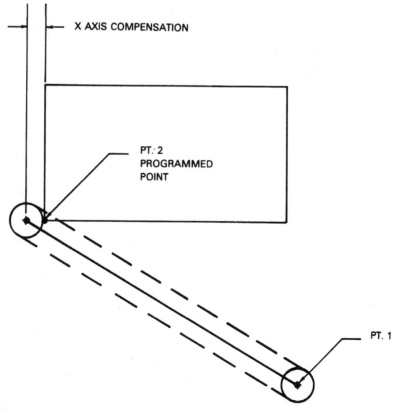

FIGURE 10-2

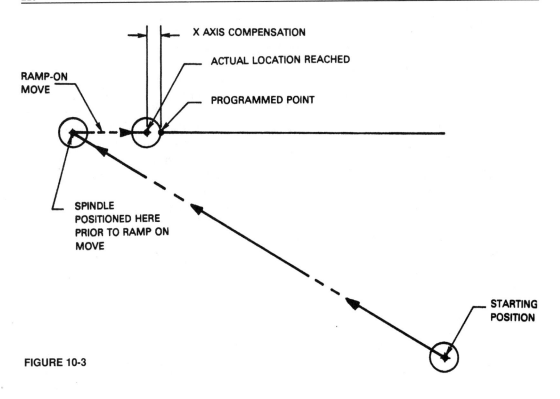

FIGURE 10-3

The ramp off move is the opposite of the ramp on move and the same precautions are necessary. In Figure 10-2, assume that cutter comp is canceled and a move is made from point #2 to point #1. In this case, the corner of the part may also be cut off. Remember, the compensation is *not turned off completely until the ramp off move is competed.*

Two additional points should be noted. First, with many controllers, *cutter comp must be* turned on after the length offset is initiated. Similarly, cutter comp must *be* cancelled prior to cancelling the tool length offset. Failure to do this will result in the controller halting executing of the program, and an alarm signaling at the MCU console.

Second, cutter comp usually must be commanded in rapid or feedrate modes, not in circular interpolation mode. If G40, G41, or G42 are commanded after G02 or G03, a machine controller alarm will result.

A short program to mill the part in Figure 10-4 is given in Figure 10-5. The program contains both a roughing and a finishing cut. One way to accomplish this without changing tools or programmed coordinates is to program the diameter of the cutter as two separate diameters. This program uses a .500-inch-diameter end mill. By defining it as both a .520-inch diameter and a .500-inch diameter and using the same coordinates for both passes, the result is that .010 inch of stock is left for the finish pass.

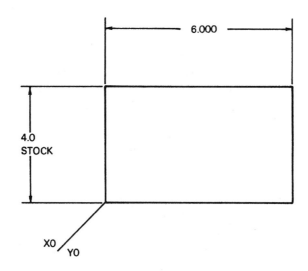

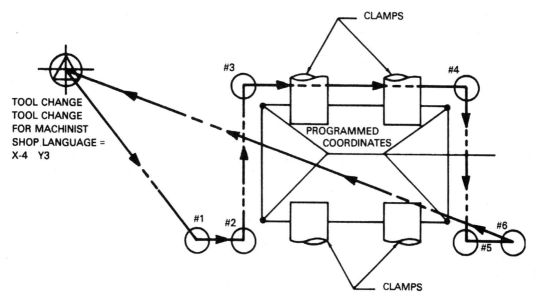

FIGURE 10-4
Part drawing and cutter path for cutter diameter compensation

```
%
O1005
(* *********)
(* X0/Y0 = LOWER LEFT CORNER)
(* CLEARANCE ABOVE CLAMPS 3.000 MIN.)
(* TOOLS = 1.000 IN. 4FLT END MILL)
(* *********)
N001G00G40G70G90
N101GG91G30X0.Y0.Z0.M19
N102T01M06
(POSITION TOOL AND PICK UP LENGTH OFFSET)
N103X-1.Y-.25S2500M03
N104G43Z0H01M08
N105G01Z-.62F12.8
(RAMP ON CRO - USE REGISTER D11)
N106G17G41X0D11
N107Y4.
N108G00Z3.
N109X6.
N110Z0.
N111G01Z-.62
N112Y0.
(RAMP OFF CRO)
N113G40X7.
N114G00Z3.
(POSITION TO START OF NEXT SEQUENCE)
N115X-1.Y-.25
N116Z0.
N117G01Z-.62
(RAMP ON CRO - USE REGISTER D12)
N118G17G41X0.D12
N119Y4.
N120G00Z3.
N121X6.
N122Z0.
N123G01Z-.62
N124Y0.
(RAMP OFF CRO)
N125G40X7.M09
N126G00Z0.M05
(CANCEL TOOL LENGTH OFFSET AND)
(RETURN TO TOOL CHANGE POSITION)
N127G91G30X0.Y0.Z0.M19
N128M30
%
```

FIGURE 10-5

Cutter diameter compensation program to mill the part in Figure 10-4

PROGRAM EXAMPLE

This program was written for a Hitachi Seiki HC-500 machining center utilizing a FANUC 11M control. The machine uses an automatic tool changer, and requires the programmer to send the spindle to the tool change position; The M06 command does not automatically do this. Additionally, the program also must explicitly orient the spindle to the correct rotation angle for the tool changer to grip the tool. To send the spindle to the tool change position, the command sequence G30G91X0.Y0.Z0. is programmed. To orient the spindle rotation, the command M19 is issued.

O1005

Program number ("O" number).

N001 – *Safety line*

G00 – Commands rapid traverse mode.
G40 – Cancels any active cutter comp.
G70 – Selects inch input.
G90 – Selects absolute positioning mode.

Blocks N101 through N104 are the tool change sequence blocks
N101

G91G30X0.Y0.Z0. – This command sends the tool to the automatic tool change position.
M19 – Orients the spindle.

N102

T01 – Places tool #1 tool change standby position.
M06 – Initiates an automatic tool change. The tool in standby position will be placed into the spindle. The tool currently in the spindle will be returned to the tool magazine.

N103

X/Y coordinates – Position the tool at location #1.
S2500 – Sets the spindle speed to 2500 rpm.
M03 – Turns the spindle on clockwise.

N104

G43 – Turns on the tool length offset compensation.
Z0 – positions the Z-axis to the desired start point.
H01 – Commands the MCU to use the value in tool register #1 for tool length compensation.
M08 – Turns on the flood coolant.

Blocks N105 through N126 are the tool motion sequence
N105

Positions the Z axis at feedrate to the proper cutting depth.

N106

This block is the cutter comp ramp on move. The cutter will be compensated at the end of this block.

G17 – Selects the X/Y plane.

G41 – Turns on cutter diameter compensation left.

X0. – Moves the spindle from location #1 to #2.

D11 – Instructs the MCU to use the values in register 11 for cutter diameter compensation.

N107

Moves the spindle from location #2 to #3 at feedrate.

N108

Rapids Z axis safely above part in preparation for jumping over the clamps.

N109

Positions the spindle at rapid from location #3 to #4.

N110

Rapids the Z axis to .100 above the part.

N111

Feeds the Z axis to milling depth.

N112

Moves the cutter from location #4 to #5 at feedrate.

N113

This is the cutter diameter compensation ramp off move. The cutter will be fully decompensated at the end of this block.

G40 – Cutter compensation cancel code.

X7. – Moves the spindle from location #5 to #6.

N114

Rapids the Z axis safely above the fixture clamps.

N115

Rapids the spindle from location #5 to #1.

N116

Rapids spindle to .100 above the part.

N117

Feeds the Z axis to the milling depth.

N118

This block is the cutter comp ramp on move. The cutter will be compensated at the end of this block.

G17 – Selects the X/Y plane.

G41 – Turns on cutter diameter compensation left.

X0. – Moves the spindle from location #1 to #2.

D12 – Instructs the MCU to use the values in register 12 for cutter diameter compensation.

N119

Moves the spindle from location #2 to #3 at feedrate.

N120

Rapids Z axis safely above part in preparation for jumping over the clamps.

N121

Positions the spindle at rapid from location #3 to #4.

N122

Rapids the Z axis to .100 above the part.

N123

Feeds the Z axis to milling depth.

N124

Moves the cutter from location #4 to #5 at feedrate.

N125

This is the cutter diameter compensation ramp off move. The cutter will be fully decompensated at the end of this block.

G40 – Cutter compensation cancel code.

X7. – Moves the spindle from location #5 to #6.

M09 – Turns off the coolant.

N126

Rapids Z axis to .100 above part and turns off the spindle.

Block N126 is the tool cancel sequence

N127

G91G30X0.Y0.Z0. – Commands the spindle at rapid traverse to the tool change position.

M19 – Orients the spindle.

N128

M30 – Signals end of program, memory reset.

SPECIAL CONSIDERATIONS

Figure 10-6 illustrates the correct method for turning compensation on or off when machining an inside pocket. Point B must be a minimum of one cutter radius away from the corner of the pocket. If point C were programmed as the ramp on move, the cutter would cut into the corner as in Figure 10-7. The direction of the cut depends on whether the X or Y axis is programmed as the first move following the G41.

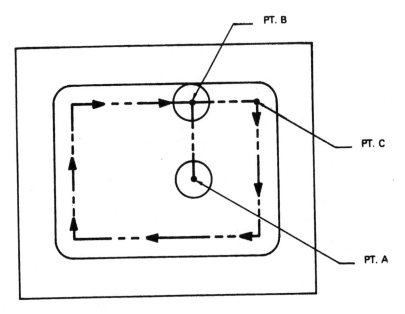

FIGURE 10-6
Turning compensation on or off when machining an inside pocket to prevent cutting into the corner

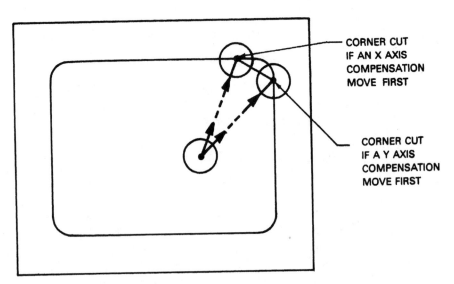

FIGURE 10-7

Figure 10-8 illustrates the precautions necessary when ramping on or off an angle. Point A should not be used for a ramp on or ramp off move since the corner of the angle will be cut off the part, and there may also be damage to the cutter. Point C, or some other point roughly perpendicular to the angle, should be used for the ramp on or ramp off move. Two different methods of positioning are used for cutter comp with respect to angles, as demonstrated in Figure 10-9. On older CNC machinery, the machine positions the cutter tangent to point A. A G code is then used to initiate the rotation from Y1 to Y2. On newer machinery, the cutter is positioned directly to point P, tangent to both line A and line Y. No special G codes are necessary in this instance. The programming manual for a particular machine will tell the programmer whether a G code is required.

Approach Angles and Vectors

Another factor to consider when using cutter diameter compensation is the approach angle used when ramping on. As Figure 10-10 illustrates, there are three possible angles that can be used during a ramp-on move: 90 degrees to the next cut, less than 90 but greater than 45 degrees to the next cut and less than 45 degrees to the next cut. Some controllers will accept any of these approach angles, others will not. If an unacceptable approach angle is used, the cutter will move to the programmed coordinates, but the cutter compensation will not take place. When programming a number of controllers, or if the NC program will be run on more than one type of controller, it is best to use a 90-degree approach angle to eliminate problems when ramping on cutter comp.

Sometimes, a controller requires a vector to be commanded with the G41 or G42 to orient the cutter correctly prior to the ramp – on move. Technically, a vector is a geometric entity that has both magnitude (length) and direction. In NC programming, vectors are simply mathematical arrows that point the cutter in a given direction. To utilize a vector the I and J addresses are used. Figure 10-11 illustrates some cutter comp vectors. If cutter comp was to be initiated from point A (Figure 10-11) and ramped on to point B, the following program blocks would be used.

For Figure 10-11 (A):

    N010 G17 G42 I-.5 J-.866 D21
    N020 G00 X6.0 Y-.5

For Figure 10-11 (B):

    N010 G17 G42 I-.866 J-.5 D21
    N020 G00 X6.0 Y-.5

For Figure 10-11 (C):

    N010 G17 G42 I-1.0 J0 D21
    N020 G00 X6.0 Y-.5

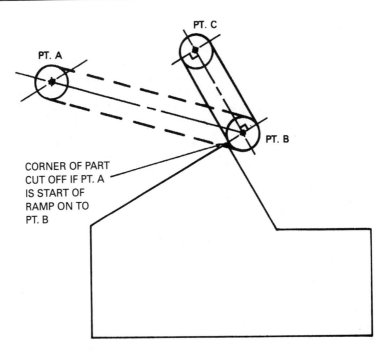

FIGURE 10-8
Ramping on or off and angle

Note that in each of these cases I is the X-axis component, and J is the Y-axis component of a vector 1.0 inch long. This is called a unit vector. In Figure 10-11 (A) the approach angle is 30 degrees, therefore, I equals the sine of 30 degrees, while J equals the cosine. In Figure 10-11 (B) the approach angle is 60 degrees. I equals the sine of 60 degrees, while J equals the cosine of 60 degrees. Since the approach angle in Figure 10-11 (C) is 90 degrees, I simply equals 1.0 and J equals zero.

Figure 10-12 shows a part to be milled using cutter diameter compensation. A program to mill the part is given in Figure 10-13. It is assumed that the part is clamped through two already existing holes. With the information given thus far, the student should be able to follow this program without further explanation.

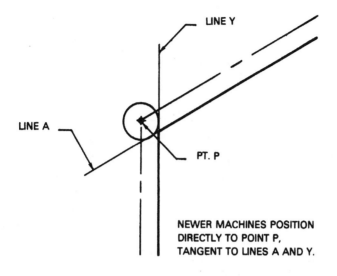

LINE Y

LINE A

PT. P

NEWER MACHINES POSITION
DIRECTLY TO POINT P,
TANGENT TO LINES A AND Y.

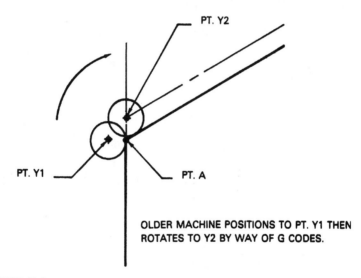

PT. Y2

PT. Y1

PT. A

OLDER MACHINE POSITIONS TO PT. Y1 THEN
ROTATES TO Y2 BY WAY OF G CODES.

FIGURE 10-9
Cutter diameter compensation of angles

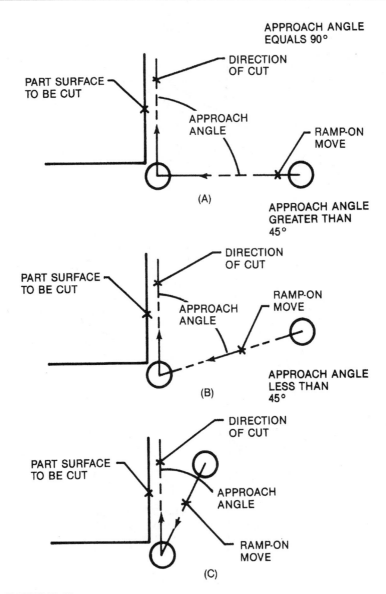

FIGURE 10-10
Cutter compensation approach angles

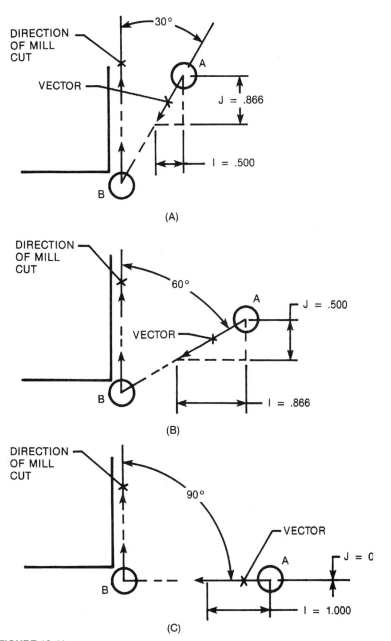

FIGURE 10-11
Cutter compensation vectors

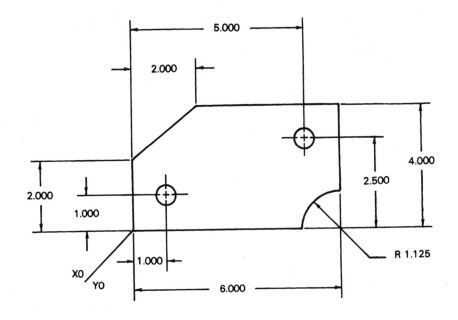

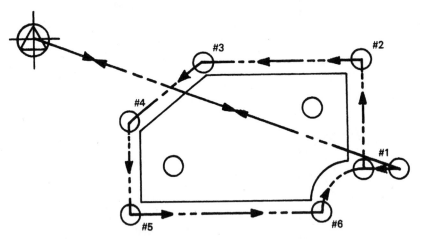

FIGURE 10-12
Part drawing and cutter path

```
%
O1013
(* *********)
(* X0/Y0 = LOWER LEFT CORNER)
(* TOOL  = 1.000 IN. 4FLT END MILL)
(* CLEARANCE ABOVE CLAMPS 3.000 MIN.)
(* *********)
N001G00G40G90G80
N100T01M06
(POSITION TOOL AND PICK UP LENGTH OFFSET)
N101G00X7.Y.875S400M03
N102G43Z0.H01M08
N103G01Z-.89F6.8
(RAMP ON CRO - USE REGISTER D11)
N104G17G42X6.D11
N105Y4.
N106X2.
N107X0.Y2.
N108Y0.
N109X4.875
N110G02X6.Y1.125I0.J1.125
(RAMP OFF CRO)
N111G00G40X7.
(RAMP ON CRO - USE REGISTER D12)
N112G17G42X6.D12
N113Y4.
N114X2.
N115X0.Y2.
N116Y0.
N117X4.875
N118G02X6.Y1.125I0.J1.125
(RAMP OFF CRO)
N119G00G40X7.5M09
N120G00Z0.M05
(CANCEL TOOL LENGTH OFFSET AND)
(RETURN TO TOOL CHANGE POSITION)
N121G91G30X0.Y0.Z0.M19
N122M30
%
```

FIGURE 10-13
Program to mill part in Figure 10-12

FINE TUNING WITH CUTTER DIAMETER COMPENSATION

Up to this point, cutter diameter compensation has been used to program the part line; the program coordinates have matched the part dimensions. Another way cutter comp is employed is to fine tune the cutter path. In this type of programming, the part is programmed using the parallel path method used in Chapters 6, 7, and 9. Cutter comp is used to compensate for the difference between the programmed and actual cutter diameter. For example, if a program is written for a .500 diameter end mill, but a resharpened end mill measuring .490 diameter is used, the .020 diameter difference can be compensated for by using cutter comp.

In the fine tune method, cutter comp is usually used to compensate for a cutter which is smaller than the programmed diameter. When using the part line method exactly the opposite is the case. Cutter comp is used to compensate for a cutter which is larger than the zero diameter cutter programmed (the part line) . For this reason it is necessary to use a minus (–) value in the cutter comp register when using the fine tune method.

Figure 10-14 is a word address program for the part in Figure 10-12, illustrating the fine tune method. Note that allowance is once again being made for the cutter radius. The cutter diameter compensation allows reground, undersize cutters to be used.

```
%
O1014
(* **********)
(* X0/Y0 = LOWER LEFT CORNER)
(* **********)
N010G00G40G70G90
N020T01M06
N030G00X7.5Y.875S400M03
N040G45Z0.H01M08
N050G01Z-.89F6.8
N060G17G42X6.51D11
N070Y4.51
N080X1.7829
N090X-.51Y2.2171
N100Y-.51
N110X5.385
N120G02X6.5Y.615I1.635J0.
N130G01Y4.5
N140X1.7929
N150X-.5Y2.2071
N160Y-.5
N170X5.375
N180G02X6.5Y.626I1.626J0.
N190G00G40X7.5M09
N200G91G30Z0.M05
N310G30X0.Y0.
N320M30
%
```

FIGURE 10-14
Program to mill the part in Figure 10-12 using the "fine tune" method

SUMMARY

The important concepts presented in this chapter are:

- Cutter diameter compensation is the automatic calculation of the cutter path by the machine control unit, based on the part line and cutter information contained in the program.
- Cutter diameter compensation is instituted and canceled through use of the codes G40, G41, and G42. G41 is cutter compensation left, G42 is cutter compensation right, and G40 is cutter compensation cancel.
- The "ramp on" move is the initial compensation of the cutter. The compensation occurs 90 degrees to the next axis movement following the G41 or G42. Care must be taken with the spindle position prior to the ramp on move to avoid cutting the part in the wrong area.
- The "ramp off" move is the opposite operation. Ramp off will occur 90 degrees to the next axis movement following a G40. The compensation will be completely eliminated by the end of this move.

VOCABULARY INTRODUCED IN THIS CHAPTER

Approach angle
Cutter diameter compensation (cutter comp.)
Cutter radius offset (CRO)
Ramp off move
Ramp on move

REVIEW QUESTIONS

1. What is cutter diameter compensation? How does it differ from tool length offset?
2. What is a ramp on move? When does it occur?
3. What is a ramp off move? When does it occur?
4. Draw a sketch illustrating the proper technique for ramping on, assuming the machine does not have the capability to compensate in two axes simultaneously. Draw a sketch illustrating an improper ramp on.
5. What cautions must be observed when instituting cutter compensation inside a pocket? When milling angles?
6. What do the codes G40, G41, and G42 do?
7. Do all CNC machines directly position the cutter with respect to an angle? If not, how is the rotation accomplished?
8. Write a program to mill the part in Figure 10-15, using a roughing and a finishing pass with a 1.000-inch-diameter end mill:

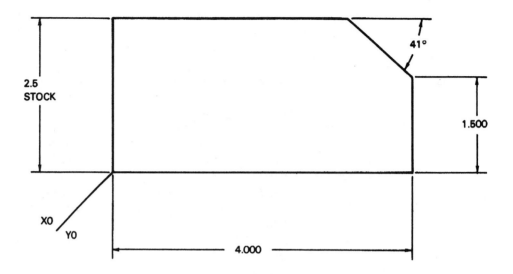

MATERIAL: 3/8 THICK 302 STAINLESS STEEL

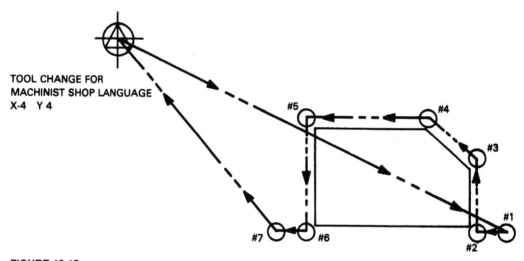

FIGURE 10-15
Part drawing for review question #8

CHAPTER 11

Do Loops and Subprograms

OBJECTIVES Upon completion of this chapter, you will be able to:

- Describe a do loop.
- Describe a subprogram.
- Describe nested loops.
- Write simple programs using loops, subroutines, and nested loops.

DO LOOPS

Figure 11-1 shows a part with a series of holes to be drilled, equally spaced. If an operation is to be repeated over a number of equal steps, it may be programmed in what is referred to as a do loop. In a *do loop*, the MCU is instructed to repeat an operation (in this case, drill a hole five times) rather than programmed for five separate hole locations.

A do loop simply instructs the MCU to repeat a series of NC program statements a specified number of times. The flowchart given in Figure 11-2 illustrates the basic construct of a do loop.

Looping capability on a CNC controller usually is an optional item, therefore not all controllers have it. The looping feature is sometimes added to the controller by the controller manufacturer. In other cases it is programmed into the controller by the machine tool manufacturer. This means that the NC codes used to initiate a do loop can vary widely from machine to machine, even if they are all equipped with the same basic controller model.

Programming a Loop

Naturally there usually is a G code to institute a do loop. As mentioned previously, there is no standard codes for do loops. The method described in this section is only one of the schemes in use. The format for a do loop is:

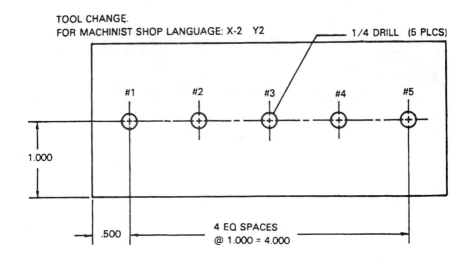

TOOL CHANGE.
FOR MACHINIST SHOP LANGUAGE: X-2   Y2   1/4 DRILL (5 PLCS)

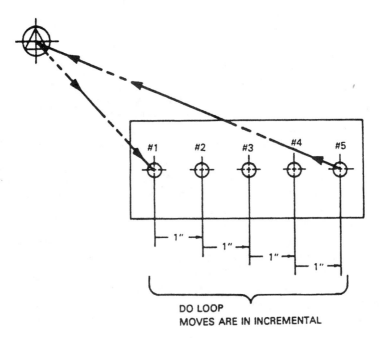

FIGURE 11-1
Part drawing and cutter path for do loop example

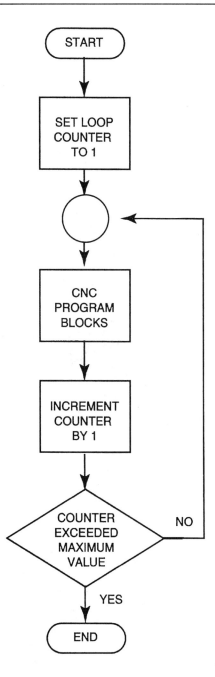

FIGURE 11-2
Process flow of a do loop

N...   G25 Pppp Qqqq Ll
Nppp X/Y/Z
N...   X/Y/Z
N...   X/Y/Z
Nqqq X/Y/Z

Where:
G25 signals the start of a loop.
P specifies the beginning block number of the loop.
Q specifies the ending block number of the loop.
L specifies the number of times to perform the loop.

The program in Figure 11-3, written to drill the part in Figure 11-1 illustrates this type of do loop used on a FANUC-style CNC controller.

```
%
O1103
(* **********)
(* X0/Y0 IS LOWER LEFT CORNER)
(* TOOL 1 - NO. 3 C-DRILL)
(* **********)
N001G00G40G80G70G8090
N101T01M06
N102G00X.5Y1.S3500M03
N103G43Z0H01M08
N104G81G99X.5Y1.Z-.162R0.F7.
(BEGIN LOOP)
N105G25P106Q106L5
N106G91X1.
N107G80M09
N108G00G91G30Z0.M05
N109G30X0.Y0.M19
N110M01
(* **********)
(* TOOL 2 - 1/4 DRILL
(* **********)
N002G00G40G80G70G8090
N201T01M06
N202G00X.5Y1.S3500M03
N203G43Z0H01M08
N204G81G99X.5Y1.Z-.375R0.F7.
(BEGIN LOOP)
N205G25P206Q206L5
N206G91X1.
N207G80M09
N208G00G91G30Z0.M05
N209G30X0.Y0.M19
N211M30
%
```

FIGURE 11-3
Program to mill the part in Figure 11-1

Program Explanation

O1103
Program number.

N001 through N103 is the tool change sequence
N001
Program safety line.

N101
T01 – Places tool 1 in the tool change standby position.

M06 – Initiates an automatic tool change.

N102
G00X.5Y1. – Positions the spindle at the X/Y starting position of the drill cycle at rapid traverse.

S3500M03 – Sets the spindle speed to 3500 rpm, and turns on the spindle in a clockwise direction.

N103
G43 – Turns on tool length compensation.

Z0. – Positions the spindle .100 above the part at rapid.

H01 – Instructs the MCU to use the value in register 1 for tool length compensation.

M08 – Turns on the coolant.

N104 through N108 are the tool motion statements
N104
G81 – Turns on the canned drill cycle.

G99 – Instructs the MCU to return the spindle to the reference plane at the end of each cycle iteration.

X.5y1. – Y/Y coordinates of the first hole.

Z-.162 – The final feed depth for the Z axis.

R0. – Sets Z0 as the reference plane (feed engagement point).

N105
G25 – Initiates the do loop. The loop counter is initialized to 1.

P106 – Tells the MCU to begin the loop in block N106.

Q106 – Tells the MCU that N106 is the end of the loop.

L5 – Instructs the MCU to perform the loop 5 times.

N106
G91 – Selects incremental positioning.

X1. – Causes the spindle to move 1.000 inch along the X axis. Since the drill cycle is still turned on, a hole will be drilled at this location.

N108
G80 – Cancels the active drill cycle.

N108 through N109 are the tool cancel commands
N108

G00G91G30Z0. – Sends the Z axis to tool change position.

M09 – Turns off the coolant.

N109

G30X0.Y0. – Sends the X and Y axes to tool change position.

M19 – Orients the spindle for tool change gripper.

N110

M01 – Optional stop code.

N002 through N203 is the tool change sequence
N002

Program safety line.

N201

T02 – Places tool 2 in the tool change standby position.

M06 – Initiates an automatic tool change.

N202

G00X.5Y1. – Positions the spindle at the X/Y starting position of the drill cycle at rapid traverse.

S3500M03 – Sets the spindle speed to 3500 rpm, and turns on the spindle in a clockwise direction.

N203

G43 – Turns on tool length compensation.

Z0. – positions the spindle .100 above the part at rapid.

H02 – Instructs the MCU to use the value in register 2 for tool length compensation.

M08 – Turns on the coolant.

N204 through N208 are the tool motion statements
N204

G81 – Turns on the canned drill cycle.

G99 – Instructs the MCU to return the spindle to the reference plane at the end of each cycle iteration.

X.5y1. – Y/Y coordinates of the first hole.

Z-.375 – The final feed depth for the Z axis.

R0. – Sets Z0 as the reference plane (feed engagement point).

N205

G25 – Initiates the do loop.

P206 – Tells the MCU to begin the loop in block N106.

Q206 – Tells the MCU that N106 is the end of the loop.

L5 – Instructs the MCU to perform the loop 5 times.

N206

G91 – Selects incremental positioning.

X1. – Causes the spindle to move 1.000 inch along the X axis. Since the drill cycle is still turned on, a hole will be drilled at this location, the do loop instructions.

N207

G80 – Cancels the active drill cycle.
M09 – Turns off the coolant.

N208 through N209 are the tool cancel commands
N208

G00G91G30Z0. – Sends the Z axis to tool change position.
M05 – Turns off the spindle.

N209

G30X0.Y0. – Sends the X and Y axes to tool change position.
M19 – Orients the spindle for tool change gripper.

N210

M30 – End of program code.

SUBPROGRAMS

A subprogram is a separate program called by another program. The use of subprograms can significantly reduce the amount of programming required on some parts. For example, on the part in Figure 11-4, note that the holes occur in the same geometric and dimensional pattern in four different locations. A do loop could be programmed to drill the holes, but programming steps can be minimized by placing the pattern in a subprogram. The drill can be sent to hole #1 and the subprogram called to drill the four holes A, B, and D. Hole #2 can then be positioned and the subprogram called again, and so on.

One way to use a subprogram is to place one or more do loops in the subprogram. This is known as *nesting*. Subprograms can also be nested in other subprograms, or nested within do loops. This gives the programmer a great deal of flexibility and a powerful programming tool.

The flowchart in Figure 11-5 illustrates how a subprogram works. The calling program is referred to as the main, or parent, program. The subprogram is sometimes referred to as the subroutine, or child program.

When the subprogram call statement is issued in the main NC program, the MCU switches to the subprogram. The subprogram then executes. At the end of the subprogram is a command that causes the MCU to switch back to the main program. The MCU returns to the NC block in the main program immediately following the command that called the subprogram in the first place.

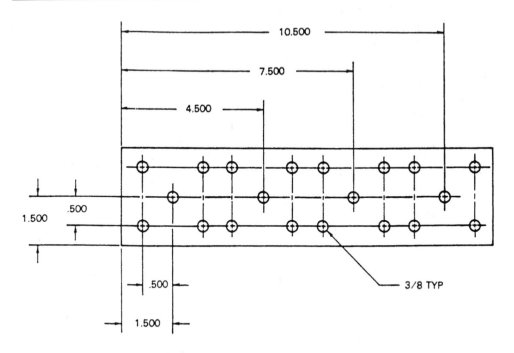

TOOL CHANGE: FOR MACHINIST SHOP LANGUAGE = X-2   Y2

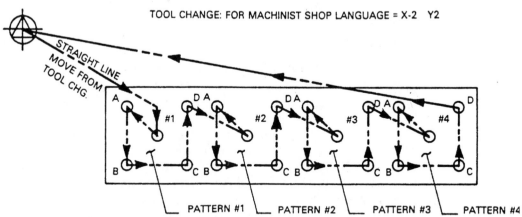

MOVES TO HOLES #1, #2, #3, #4 IN ABSOLUTE

MOVES TO HOLES A, B, C, D IN SUBROUTINE ARE INCREMENTAL

FIGURE 11-4
Part drawing and tool path for subprogram example

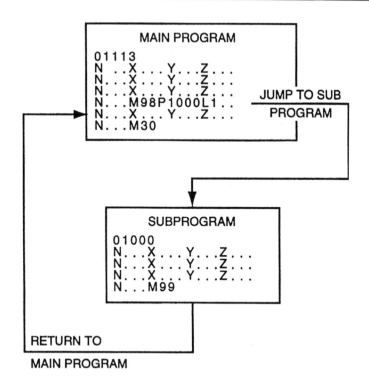

FIGURE 11-5
Subprogram call mechanics

CALLING A SUBPROGRAM

On FANUC and FANUC-style controllers, a subprogram is a program in its own right. It has its own "O" number to identify it, and is sequence numbered independently of the parent program.

The format for calling a subprogram is

MAIN PROGRAM

O0001
N001 X/Y/Z
N002   .
N003   .
N004 M98P2000L1
N005   .
N006   .
N007   .
N008 M30

SUBPROGRAM

O2000
N001 X/Y/Z
N002   .
N003   .
N004 M99

Where
M98 – Instructs the MCU to jump to a subprogram.
P2000 – Tells the MCU that O2000 is the subprogram ID.
L1 – Instructs the MCU to execute the subprogram one time.

The program in Figure 11-6 illustrates how a subprogram can be used to drill the part in Figure 11-4.

Program Explanation

O1106
Program id number.

N001 through N103 is the tool change sequence
N001
Safety block.

N101
T01 – Put tool 1 in tool change standby position.
M06 – Initiate a tool change. Tool 1 will be placed in the spindle.

N102
G00G90X1.5Y1.5 – Move to hole #1 at rapid, using absolute positioning.
S3500M03 – Turns on the spindle clockwise at 3500 rpm.

N103
G43 – Turns on tool length compensation.
Z3. – Positions the spindle to Z3.000
H01 – Instructs the MCU to use the values in offset register 1 for tool length compensation.
M08 – Turns on the flood coolant.

N104 through N112 are the tool motion blocks
N104
G81 – Turns on the canned drilling cycle.
G99 – Specifies a retract to the reference plane at the end of each drill cycle iteration.
X1.5Y1.5 – X/Y coordinates of the 1st hole.
Z-.162 – Final drilling depth.
R0. – The reference plane. This is the Z coordinate where the MCU will begin feeding the tool into the workpiece.
F10.5 – Specifies a drilling feedrate of 10.5 inches per minute.

```
%
O1106
(* **********)
(* X0/Y0 - LOWER LEFT CORNER)
(* Z0 - .100 ABOVE TOP OF PART)
(* **********)
(* TOOL 1 - NO. 3 C-DRILL)
(* **********)
N001G17G90G40G98
N101T01M06
N102G00G90X1.5Y1.5S3500M03
N103G43G90Z3.H01M08
N104G81G99X1.5Y1.5Z-.162R0.F10.5
N105P1000M98L1
N106X4.5Y1.5
N107P1000M98L1
N108X7.5Y1.5
N109P1000M98L1
N110X10.5Y1.5
N111P1000M98L1
N112G80M09
N113G00G91G28Z0.M05
N114G28X0.Y0.M01
(* **********)
(* TOOL 1 - NO. 3 C-DRILL)
(* **********)
N002G90G40G98
N201T02M06
N202G00X1.5Y1.5S3500M03
N203G43G90Z3.H01M08
N204G81G99X1.5Y1.5Z-.375R0.F10.5
N205P1000M98L1
N206X4.5Y1.5
N207P1000M98L1
N208X7.5Y1.5
N209P1000M98L1
N210X10.5Y1.5
N211P1000M98L1
N212G80M09
N213G00G91G28Z0.M05
N214G28X0.Y0.
N215M30


O1000
(* **********)
(* START OF SUBPROGRAM 1000)
(* **********)
N001G91X-.5Y.5
N002Y-1.
N003X1.
N004Y1.
N004G90
N006M99
%
```

FIGURE 11-6

Program to machine the part in Figure 11-4

N105

This block is a jump to subprogram call.

P1000 – Specifies program O1000 as the target of the subroutine jump.

M98 – Instructs the MCU to jump to the target program (in this case O1000).

L1 – Tells the MCU to execute the subprogram 1 time.

N106

X4.5Y1.5 – Positions the spindle over hole #2.

N108

X7.5Y1.5 – Positions the spindle over hole #3.

N109

P100M98L1 – Jump to subprogram O1000 call.

N110

X10.5Y1.5 – Positions the spindle over hole #4.

N111

P100M98L1 – Jump to subprogram O1000 call.

N112

G80 – Cancels the canned drilling cycle.

M09 – Turns off the flood coolant.

N113 through N114 are the tool cancel sequence
N113

G00G91G28Z0. – Sends the spindle to the Z axis home position in rapid traverse mode.

M05 – Turns off the spindle.

N114

G28X0.Y0. – Sends the spindle to the X/Y home position.

M01 – Optional program stop code. Provided for operator convenience.

N002 through N203 is the tool change sequence
N002

Safety block.

N201

T02 – Put tool 2 in tool change standby position.

M06 – Initiate a tool change. Tool 2 will be placed in the spindle.

N202

G00G90X1.5Y1.5 – Move to hole #1 at rapid, using absolute positioning.

S3500M03 – Turns on the spindle clockwise at 3500 rpm.

N203

G43 – Turns on tool length compensation.

Z3. – Positions the spindle to Z3.000

H02 – Instructs the MCU to use the values in offset register 2 for tool length compensation.

M08 – Turns on the flood coolant.

N204 through N212 are the tool motion blocks
N204

G81 – Turns on the canned drilling cycle.

G99 – Specifies a retract to the reference plane at the end of each drill cycle iteration.

X1.5Y1.5 – X/Y coordinates of the 1st hole.

Z-.375 – Final drilling depth.

R0. – The reference plane. This is the Z coordinate where the MCU will begin feeding the tool into the workpiece.

F10.5 – Specifies a drilling feedrate of 10.5 inches per minute.

N205

P100M98L1 – Jump to subprogram O1000 call.

N206

X4.5Y1.5 – Positions the spindle over hole #2.

N208

X7.5Y1.5 – Positions the spindle over hole #3.

N209

P100M98L1 – Jump to subprogram O1000 call.

N210

X10.5Y1.5 – Positions the spindle over hole #4.

N211

P100M98L1 – Jump to subprogram O1000 call.

N212

G80 – Cancels the canned drilling cycle.

M09 – Turns off the flood coolant.

N213 through N214 are the tool cancel sequence
N213

G00G91G28Z0. – Sends the spindle to the Z axis home position in rapid traverse mode.

M05 – Turns off the spindle.

N214

G28X0.Y0. – Sends the spindle to the X/Y home position.

N214

M30 – End of program memory reset code.

Subprogram Explanation

Notice that a subprogram has its own program id number, in this case O1000. The sequence blocks also are numbered independently from the main program. The only difference between the subprogram and an independent program is the return to calling program command (M99) at the end of the program.

O1000
Program id number.

N001
G91 – Selects incremental positioning mode. Incremental moves are used throughout the subprogram.

X-.5Y.5 – Moves the spindle to hole A. Since the drill cycle was turned on in the main program, a hole will be drilled at this location.

N002
Y-1. – Moves the spindle from hole A to hole B.

N003
X1. – Moves the spindle from hole B to hole C.

N004
Y1. – Moves the spindle from hole C to hole D.

N005
G90 – Selects absolute positioning mode. Since the main program uses absolute mode, operator confusion is minimized by placing the machine back into absolute mode before returning from the subprogram.

N006
M99 – Return to calling program code. The MCU will return to the main program when this code is received. If this were an independent program, not a subprogram, an M30 would have been used here instead of M99.

SUBROUTINES FOR CUTTER DIAMETER COMPENSATION

In chapter 10 it was pointed out that subprograms often are used with cutter diameter compensation. Figure 11-7 presents a program written for the part shown in Chapter 10, Figure 10-12. The program utilizes a subprogram to mill the part periphery.

The first time the subprogram is called, CRO register D11 is used. If .520 is placed in the register, .010 stock per side of the part will be left for finishing. The second time the subprogram is called, register D11 is used. Placing a

```
%
O1107
(* **********)
(* X0/Y0 - LOWER LEFT CORNER)
(* Z0 - .100 ABOVE TOP OF PART)
(* THIS PROGRAM CALLS SUBPROGRAM 1007)
(* **********)
N001G00G40G70G90        (SAFETY LINE)
N101T01M06              (TOOL CHANGE)
N102G00X7.Y.875         (MOVE CUTTER TO CRO RAMP-ON POSITION)
N103G45Z0H01M08         (PICK UP LENGTH OFFSET/COOLANT ON)
N104G01Z-.89            (POSITION CUTTER TO DEPTH)
N105G17G42D11           (TURN ON THE CRO - USE REGISTER 11)
N106P1007M98L1          (JUMP TO SUBROUTINE TO MILL PART)
N107G00X7.Y.875         (POSITION THE CUTTER FOR CRO RAMP-ON)
N108G01Z-.89            (POSITION CUTTER TO DEPTH)
N109G17G42              (TURN ON THE CRO - USE REGISTER 12)
N110P1007M98L1          (JUMP TO SUBROUTINE TO MILL PART)
N111G00G91G28Z0.M09     (RAPID Z TO HOME/COOLANT OFF)
N112G28X0.Y0.M05        (RAPID X/Y TO HOME/SPINDLE OFF)
N113M30                 (END OF PROGRAM)


O1007
(* **********)
(* SUBPROGRAM CALLED BY PROGRAM 1107)
(* **********)
N001G01X6.                    (CRO ALREADY ON - RAMP-ON TO#1))
N002Y4.                       (FEED TO #2)
N003X2.                       (FEED TO #3)
N004X0.Y2.                    (FEED TO #4)
N005Y0.                       (FEED TO #5)
N006X4.875                    (FEED TO #6)
N007G02X6.Y1.125I0.J1.125     (CUT ARC FROM #6 TO #1)
N008G40X7.                    (CANCEL CRO AND RAMP-OFF)
N009G00Z0.                    (RETRACT THE SPINDLE)
N010M99                       (RETURN TO CALLING PROGRAM)
%
```

FIGURE 11-7
Program to machine the part in Figure 10-12

value of .500 in D12 results in a finish pass. Remark statements are included in the program listing which explain the program execution in detail.

The machining sequence used in this program is identical to that used in Chapter 10. The difference is that the duplication of coordinate locations is eliminated by using the subprogram.

NESTED LOOPS

Do loops may nest inside other do loop or subprograms. Similarly, subprograms may nest inside other subprograms. This concept will be demonstrated using the part illustrated in Figure 11-8 and the program in Figure 11-9. This program features two loops nested inside a subprogram. In Figure 11-8, the rows of holes have been labeled for easy reference. In writing a CNC program, a reference sketch such as this is a valuable aid in developing a machining strategy and provides a way for the programmer to check his or her work.

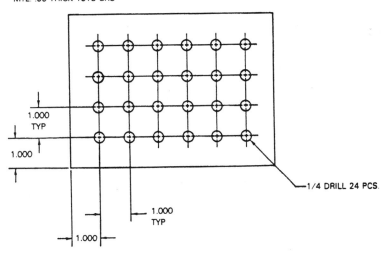

MTL: .50 THICK 1018 CRS

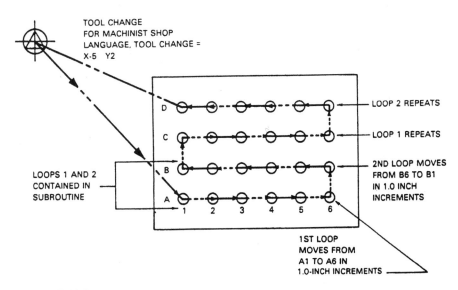

FIGURE 11-8
Part drawing and tool path for nested loop example

```
%
O1109
(* **********)
(* X0/Y0 - LOWER LEFT CORNER)
(* Z0 - .100 ABOVE TOP OF PART)
(* **********)
(* TOOL 1 -  NO. 3 C-DRILL)
(* **********)
N001G00G90G80G40                  (SAFETY LINE)
N101T01M06                        (TOOL CHANGE)
N102G00X1.Y1.S1700M03             (POSITION AT A-1)
N103G81G98X1.Y1.Z-.162R0.F5.1     (TURN ON DRILL CYCLE - DRILL A-1)
N105P1000M98L1                    (JUMP TO SUBPROGRAM 1000)
N106Y1.                           (POSITION TO AND DRILL C-1)
N107P1000M98L1                    (JUMP TO SUBPROGRAM 1000)
N108G00G91G28Z0.M09               (RAPID Z HOME/COOLANT OFF)
N109G28X0.Y0.M05                  (RAPID X/Y HOME/SPINDLE OFF)
N110M01                           (OPTIONAL STOP CODE)
(* **********)
(* TOOL 2 -  1/4 DRILL)
(* **********)
N002G00G90G80G40                  (SAFETY LINE)
N201T01M06                        (TOOL CHANGE)
N202G00X1.Y1.S1700M03             (POSITION AT A-1)
N203G81G98X1.Y1.Z-.162R0.F5.1     (TURN ON DRILL CYCLE- DRILL A-1)
N205P1000M98L1                    (JUMP TO SUBPROGRAM 1000)
N206Y1.                           (POSITION TO AND DRILL C-1)
N207P1000M98L1                    (JUMP TO SUBPROGRAM 1000)
N208G00G91G28Z0.M09               (RAPID Z HOME/COOLANT OFF)
N209G28X0.Y0.M05                  (RAPID X/Y HOME/SPINDLE OFF)
N210M30                           (END OF PROGRAM - MEMORY RESET)


O1000
(* **********)
(* SUBPROGRAM 1000 CALLED BY)
(* PROGRAM NUMBER 1109)
(* **********)
N101G25P112Q102L5                 (BEGIN DO LOOP - REPEAT 5 TIMES)
N102G91X1.                        (MOVE INCREMENTALLY 1.000 IN X)
N103Y1.                           (POSITION TO C-1)
N104G25P104Q104L5                 (BEGIN A LOOP - REPEAT 5 TIMES)
N105G91X-1.                       (MOVE INCREMENTALLY -1.000 IN X)
N106G90                           (SWITCH TO ABSOLUTE POSITIONING)
N107M99                           (RETURN TO CALLING PROGRAM)
%
```

FIGURE 11-9

Program to machine the part in Figure 11-8

In the part program, one of the loops will drill the holes in row A, moving from hole A1 to hole A6. Hole A1 will be drilled prior to instituting the do loop. After positioning to drill hole B6, another loop will be used to drill holes B6 to B1, moving in the −X direction. One loop could drill all the holes, but that would require sending the spindle back to the first hole of each row prior to using the do loop. By using two do loops, machine motion is more efficient, drilling in both the positive and negative directions along the X axis. Nesting the loops in a subprogram allows drilling rows C and D with the same do loops.

SUMMARY

The important concepts presented in this chapter are

- A do loop instructs the MCU to repeat a series of instructions a specified number of times.
- The format for a do loop is:

G25 P... Q... L..

Where: G25 turns on the loop.
P is the beginning block number of the loop.
Q is the ending block number of the loop.
L is the number of times to repeat the loop.

- A subprogram is a program called by another program in a parent-child relationship.
- The format for calling a subprogram is:

P.... M98 L..

Where P is the program number of the subprogram.
M98 causes subprogram P to execute.
L specifies the number of times subprogram P executes.

- Nested loops are placed inside other loops or inside subprograms.
- The codes for subprograms and do loops vary from controller to controller. To program a particular machine, it will be necessary to consult the programming manual for the machine in question.

VOCABULARY INTRODUCED IN THIS CHAPTER

Do loop
Main program
Nested loop
Subprogram
Subroutine

REVIEW QUESTIONS

1. What is a do loop? A subprogram? A nested loop?
2. What is the format for a do loop?
3. What is the format for a subprogram call?
4. Write a do loop to drill the hole patterns in Figure 11-10.

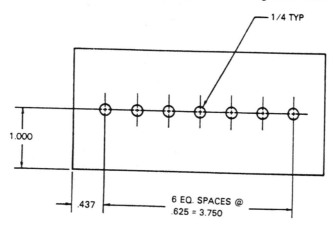

1/4 TYP

1.000

.437

6 EQ. SPACES @
.625 = 3.750

MATERIAL: 1/8 THICK 2024 T-3 ALUMINUM

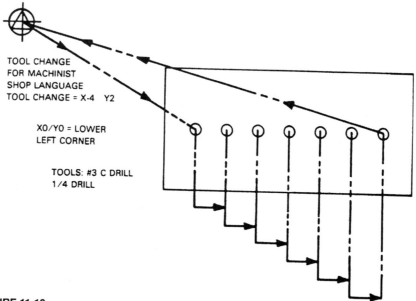

TOOL CHANGE
FOR MACHINIST
SHOP LANGUAGE
TOOL CHANGE = X-4   Y2

X0/Y0 = LOWER
LEFT CORNER

TOOLS: #3 C DRILL
1/4 DRILL

FIGURE 11-10
Part drawing for review question #4

5. Write a program utilizing a subprogram to mill the slots in Figure 11-11.

MTL 2075 ALUM. ALLOY .25 THICK

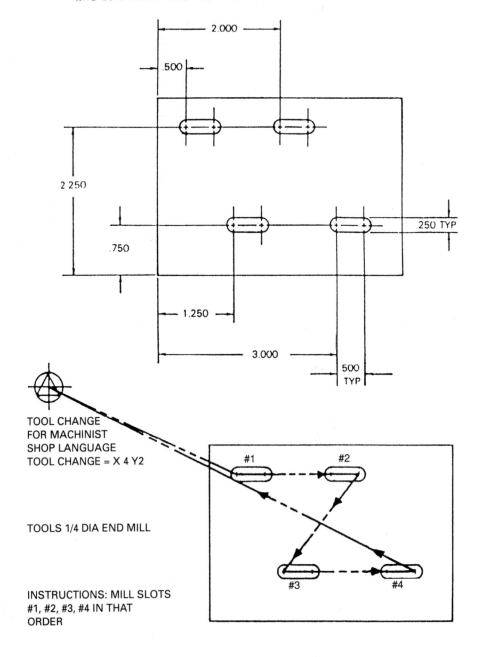

TOOL CHANGE
FOR MACHINIST
SHOP LANGUAGE
TOOL CHANGE = X 4 Y2

TOOLS 1/4 DIA END MILL

INSTRUCTIONS: MILL SLOTS
#1, #2, #3, #4 IN THAT
ORDER

FIGURE 11-11
Part drawing for review question #5

CHAPTER 12

Advanced CNC Features

OBJECTIVES Upon completion of this chapter, you will be able to:

- Explain the concept of mirror imaging.
- Decide when the use of mirror imaging is appropriate.
- Write simple programs in word address that employ mirror imaging.
- Explain the concept of polar rotation.
- Decide when the use of polar rotation is appropriate.
- Write simple programs in word address that employ polar rotation.
- Write simple programs in word address that employ polar rotation used in a do loop.
- Explain the concept of helical interpolation.
- Decide when the use of helical interpolation is appropriate.
- Write simple programs in word address that employ helical interpolation.

MIRROR IMAGING

Mirror imaging is a simple concept that can be very useful in programming. In essence, *mirror imaging* reverses the sign (+ or −) of an axis direction. For example, mirror imaging can be employed to shorten the amount of programming required to make the part shown in Figures 12-1 and 12-2. Calling the centerline of this part X0/Y0, the pattern of holes to the right of the centerline can be programmed in a subroutine. After this pattern is drilled, mirror imaging along the X axis can be instituted and the subroutine called again. This will drill the same pattern of holes in the second quadrant, with no additional programming save the mirror imaging command. The process can be repeated, mirror imaging the Y axis to drill the pattern in the third quadrant. Canceling the mirror image on the X axis and leaving it active on Y will drill the pattern in the fourth quadrant.

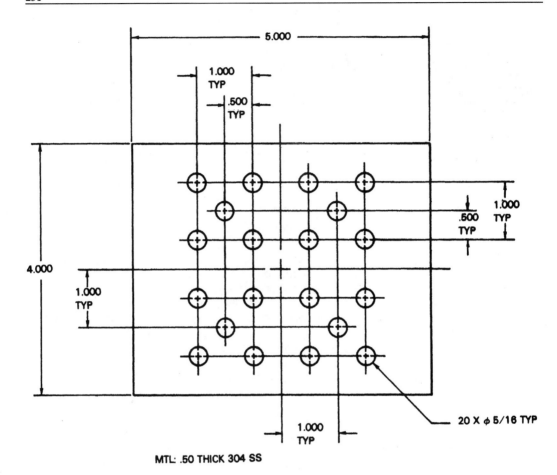

FIGURE 12-1
Part drawing

Word Address Format

In word address, this procedure is accomplished through either G codes or M functions, depending on the controller. In the following example, M functions are used as follows:

M21—Mirror image X axis.
M22—Mirror image Y axis.
M23—Mirror image off.

On some CNC machines, mirror imaging is selected at the MDI console by means of a switch. When programming such a machine, a dwell must be programmed at the place where mirror imaging is to be instituted, and instructions given for the operator to set the switches prior to restarting the program. The program to drill the part is shown in Figure 12-3.

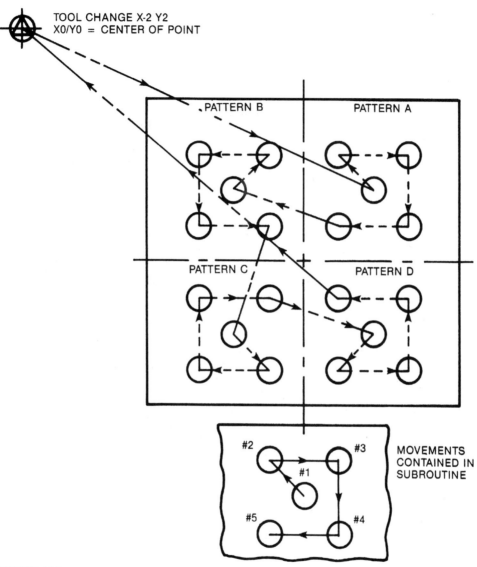

TOOL CHANGE X-2 Y2
X0/Y0 = CENTER OF POINT

PATTERN B   PATTERN A

PATTERN C   PATTERN D

#2   #3   MOVEMENTS
#1   CONTAINED IN
                 SUBROUTINE
#5   #4

FIGURE 12-2
Tool path for the part shown in Figure 12-1

```
%
O1203
(* *********)
(* X0/Y0 = CENTER OF PART)
(* THIS PROGRAM CALLS SUBPROGRAM 1000)
(* **********)
N001G00G40G90G70              (SAFETY LINE)
N101G00X0.Y0.S641M03          (POSITION OVER PART CENTER)
N102G45Z0H01M08               (PICK UP LENGTH OFFSET)
N103M98P1000L1                (CALL SUBPROGRAM)
N104M21                       (MIRROR X-AXIS)
N105M98P1000L1                (CALL SUBPROGRAM)
N106M22                       (MIRROR Y-AXIS)
N107M98P1000L1                (CALL SUBPROGRAM)
N108M23                       (TURN OFF AXIS MIRROR)
N109M22                       (MIRROR Y-AXIS)
N110M98P1000L1                (CALL SUBPROGRAM)
N111G91G28Z0.M09              (Z AXIS TO HOME ZERO)
N112G28X0.Y0.M05              (X/Y TO HOME ZERO)
N113M30                       (END OF PROGRAM)

O1000
(* **********)
(* SUBPROGRAM FOR PROGRAM 1203)
(* **********)
N001G81G90G99X1.Y1.Z-.7R0.F5. (DRILL CYCLE - DRILL HOLE 1)
N002G91X-.5Y-.5               (DRILL HOLE 2)
N003X1.                       (DRILL HOLE 3)
N004Y-1.                      (DRILL HOLE 4)
N005X-1.                      (DRILL HOLE 5)
N006G80G90                    (CANCEL DRILL CYCLE)
N007M99                       (RETURN TO MAIN PROGRAM)
%
```

FIGURE 12-3
Program to machine part in Figure 12-1

Main Program Explanation

N001
Safety line.

N101
G00X0.Y0. – Positions spindle over the center of the part.
S641M03 – Turns on the spindle clockwise at 641 rpm.

N102
G45Z0.H01 – Turns on tool length offset using register 1.
M08 – Turns on the coolant.

N103

M98P1000L1 – Executes subprogram 1000 one time. The part located in the first quadrant will be drilled.

N104

M21 – Turns on mirror imaging on the X-axis.

N105

M98P1000L1 – Executes subprogram 1000 one time. Since the X axis is mirrored, the part in the second quadrant will be drilled.

N106

M22 – Turns on mirror imaging on the Y axis.

N107

M98P1000L1 – Executes subprogram 1000 one time. Both the X and Y axes are mirrored. Therefore, the part in the third quadrant will be drilled.

N108

M23 – Turns off all mirror imaging.

N109

M22 – Turns the Y axis mirror image on. Only the Y axis is now mirrored.

N110

M98P1000L1 – Executes subprogram 1000 one time. The part in the fourth quadrant will now be drilled.

N111

G91G28Z0. – Returns the Z axis to the home zero position.
M09 – Turns off the coolant.

N112

G28X0.Y0. – Returns the X and Y axes to the home zero position.

N113

M30 – End of program code.

Subprogram Explanation

N001

G81 – Turns on the drill cycle.
G90 – Places the MCU in absolute positioning mode.
G99 – Specifies a return to reference level when the Z axis retracts after each cycle iteration.
X1.Y1. – Coordinates of hole #1.
Z-.7 – The final drill depth.
R0. – Specifies Z0. as the feed engagement point.
F5. – Sets the drilling feedrate at 5.0 inches per minute.

N002

G91 – Places the MCU in incremental positioning mode.

X-.5Y-.5 – Coordinates of hole #2.

N003

X1. – Coordinates of hole #3. The MCU is still in incremental mode.

N004

X-1. – Coordinate of hole #4.

N005

M99 – Returns the MCU to the main program.

POLAR ROTATION

Consider the part shown in Figure 12-4, in which four slots are to be milled. A machinist making this part on a conventional vertical milling machine would probably set up the workpiece on a rotary table, rotate 45 degrees from the nominal 0-degree location, and mill the first slot. The other three slots could then be milled, moving the various axes, or the machinist could simply index the part 90 degrees from the first slot to mill the second without excess movement along the X and Y axes. The same type of machining may be accomplished on a CNC machining center or CNC mill equipped with polar rotation.

A polar axis coordinate system is formed by constructing a line whose slope is not the same as either the X or Y axis. For example, in Figure 12-5, a line has been constructed between the origin (point #1) and point #2 on the graph. That line is a polar axis. Notice that point #2 is located 1.0 inch from the origin as measured along the polar axis. If point #2 is specified as (1,0) measured along the polar axis, then point #2 is called a polar coordinate. In mathematics more scientific definitions exist for a polar axis, but for the purposes of CNC programming, *polar rotation* can be thought of as rotating the Cartesian coordinate system.

When polar rotation is instituted in a CNC program, the MCU will triangulate the points necessary to position the tool to the desired coordinates from the program information that it is given. Polar rotation is supplied on most controllers as an optional feature. As with most options, the coding for polar rotation varies greatly from machine to machine. The examples given here can serve only to demonstrate the concept. The NC part programmer will have to consult the programming manual to program polar rotation successfully on a given machine.

Despite the differences in controllers, there is certain information that every MCU needs in order to carry out a polar rotation:

• The X axis coordinate of the center of rotation.
• The Y axis coordinate of the center of rotation.

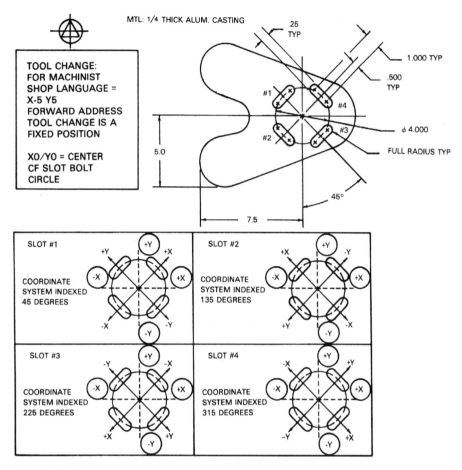

MTL: 1/4 THICK ALUM. CASTING

TOOL CHANGE:
FOR MACHINIST
SHOP LANGUAGE =
X-5 Y5
FORWARD ADDRESS
TOOL CHANGE IS A
FIXED POSITION

XO/YO = CENTER
CF SLOT BOLT
CIRCLE

FIGURE 12-4
Part drawing

- The *index angle,* or the angle as measured counterclockwise from the + X axis to the start of the rotation. In the case shown in Figure 12-4, the index angle is 45 degrees. This value is the angular rotation from the X axis to slot #1.
- The amount of the rotation. Following the initial rotation to the index angle, subsequent rotations may be specified as some angular value other than the index angle. The rotations will occur in a counterclockwise direction. In the case shown in Figure 12-4, this amount is 90 degrees. In other words, following the initial index of the coordinate system 45 degrees, subsequent rotations will be 90 degrees until the cancel command is given.
- A code to initiate polar rotation.
- A code to cancel polar rotation.

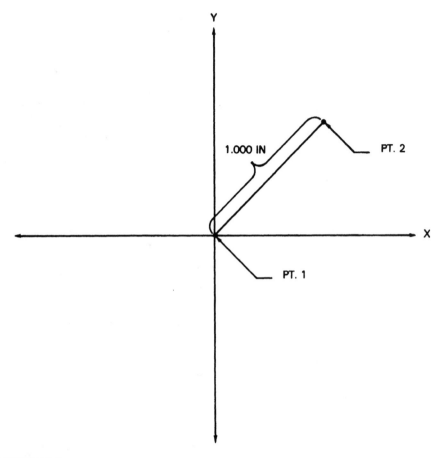

FIGURE 12-5

Word Address Format

The program to mill the slots is presented in Figure 12-6. The coding format here is designed to be generic for the purposes of instruction. Every controller uses a different coding method for polar rotations, and many controllers do not offer the capability. Polar rotation is used generally on three-axis machinery to compensate for the lack of a fourth rotary axis. The format for word address polar rotations used in this book is:

G61 X. . . .Y. . . . A . . . D . . . L . .
Programming information
G60

Where G61 is the code to institute polar rotation, X. . . . is the X axis center of rotation, Y. . . . is the Y axis center of rotation, A. . . . is the index angle measured in degrees from the X axis, D. . . . is the subsequent amount of rotation measured in degrees, L. . . . is the number of rotations to be performed, and G60 is the code to cancel the rotation.

```
%
O1206
(* **********)
(* X0/Y0 = CENTER OF PART)
(* CALLS SUBPROGRAM 2000)
(* **********)
N001G00G40G90G70
N101G00X0.Y0.S3500M03        (INITIAL MOVE TO CENTER OF PART)
N102G45Z0.H01M08             (PICK UP LENGTH OFFSET)
N103G61X0.Y0.A45D90L4        (INVOKE POLAR ROTATION)
N104M98P2000L1               (CALL SUBPROGRAM)
N105G61                      (INVOKE POLAR ROTATION)
N106M98P2000L1               (CALL SUBPROGRAM)
N107G61                      (INVOKE POLAR ROTATION)
N108M98P2000L1               (CALL SUBPROGRAM)
N109G61                      (INVOKE POLAR ROTATION)
N110M98P2000L1               (INVOKE SUBPROGRAM)
N111G60                      (CANCEL POLAR ROTATION)
N112G00G91G28Z0.M09          (RETRACT Z TO HOME ZERO)
N113G28X0.Y0.M05             (X/Y TO HOME ZERO)
N114M30                      (END OF PROGRAM


O2000
(* **********)
(* SUBPROGRAM FOR PROGRAM 1206)
(* **********)
N001G00X-.5Y2.               (POSITION TO THE SLOT)
N002G01Z-.36F2.8             (FEED TO MILLING DEPTH)
N003X.5                      (MILL SLOT)
N004G00Z0.                   (RAISE THE SPINDLE)
N005M99                      (RETURN TO MAIN PROGRAM)
%
```

FIGURE 12-6
Program to machine part in Figure 12-4

Main Program Explanation

N001
Safety line.

N101
G00X0.Y0. – Positions the spindle at the center of the part.
S3500M03 – Turns on the spindle clockwise at 3500 rpm.

N102
G45Z0.H01 – Turns on the tool length offset compensation.
M08 – Turns on the coolant.

N103

G61 – Polar rotation on code.
X0.Y0. – Center point of the rotation.
A45 – Starting index angle of rotation.
D90 – Amount to rotate coordinate system with each G61 call.
L4 – Tells the MCU that 4 rotations will occur.

N104

M98P2000L1 – Jumps to subprogram O2000.

N105

G61 – Causes the next polar rotation to occur.

N106

M98P2000L1 – Jumps to subprogram O2000.

N107

G61 – Causes the next polar rotation to occur.

N108

M98P2000L1 – Jumps to subprogram O2000.

N109

G61 – Causes the next polar rotation to occur.

N110

M98P2000L1 – Jumps to subprogram O2000.

N111

G60 – Turns off the polar rotation.

N112

G00G91G28Z0. – Sends the Z axis to home zero.
M09 – Turns off the coolant.

N113

G28X0.Y0. – Sends the X/Y axes to home zero.
M05 – Turns off the spindle.

N114

M30 – End of program code.

Subprogram Explanation

N001

G00X-.5Y2. – Positions spindle to the start of the slot.

N002

G01Z-.36F2.8 – Feeds the spindle to milling depth at a feedrate of 2.8 inches per minute.

N003

X.5 – Positions the spindle at feedrate to the end of the slot.

N004

G00Z0. – Retracts the cutter from the slot.

N005

M99 – Returns the MCU to the main program.

HELICAL INTERPOLATION

Helical interpolation is another useful feature of CNC machinery. *Helical interpolation* allows circular interpolation to take place in two axes (usually X and Y), while subsequently feeding linearly with the third (usually Z). This makes possible the milling of helical pockets and threads.

Figure 12-7 shows a part on which a 1.000-20 thread is to be machined. An oddly shaped part like this can be cut on a CNC machine as easily as setting it up on a face plate on a lathe or a four-jaw chuck. In the case of a production run, machining this part on the mill eliminates the need for extra fixturing. It can be threaded in the same setup used to mill it to shape. The programs presented here will assume that the part has been cast separately, however, leaving only the thread to be milled. The thread will be cut by circular interpolation with the X and Y axes, while feeding with the Z axis.

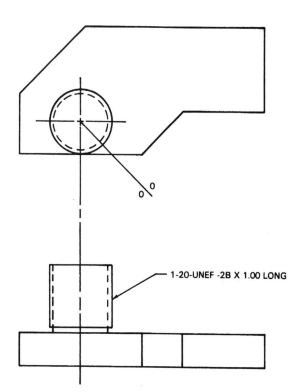

1-20-UNEF -2B X 1.00 LONG

FIGURE 12-7
Part drawing

A special type of milling cutter, called a thread hob, will be used to mill the thread. The hob will be sent to a start position, fed into the workpiece, then helically interpolated for three turns. The cutter will then be withdrawn from the part. With each turn the hob makes around the part, the Z axis will advance downward an amount equal to the lead of the thread. The lead of the thread can be determined by the formula:

$$L = P \times I$$

Where L is the lead of the thread, P is the pitch of the thread, and I is the number of leads on the thread. The pitch of a thread is 1 divided by N, where N is the number of threads per inch. For a 20 thread, the pitch is 1 divided by 20 or .050 inch. The lead for a single lead 20 thread is .050 times 1, or .050. This means that the thread will advance .050 inch in one revolution. Note that the value of the lead and the pitch on a single lead thread are identical; however, the lead and pitch are not the same thing.

A 60-degree thread milling cutter is used to mill the thread on the part in Figure 12-7. Set up as shown in Figure 12-8.

Program Format

Not every CNC machine has helical interpolation. It is usually an optional feature, purchased at additional cost. Helical interpolation in word address can be accomplished in any one of three plane combinations (X/Y, Z/X, or Y/Z). To select the planes, the following G codes are used:

G17—X/Y plane.
G18—X/Z plane.
G19—Y/Z plane.

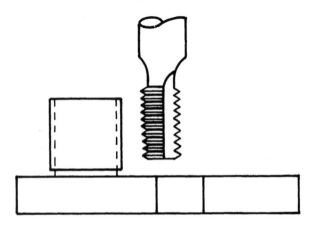

FIGURE 12-8
Thread milling cutter setup

The format for helical interpolation in word address is as follows:

- *For the X/Y plane*—G17 G02/G03 X. . . . Y.... I. . . . J. . . . Z. . . . F. . .
- *For the X/Z plane*—G18 G02/G03 X. . . . Y....I. . . . K. . . . Z. . . . F. . .
- *For the Y/Z plane*—G19 G02/G03 X. . . . Y.... J. . . . K. . . . Z. . . . F. . .

Where: G17, G18, and G19 select the plane, G02 and G03 select the direction of helical interpolation (G02 clockwise, G03 counterclockwise), X, Y, and Z are the arc endpoint coordinates, I, J, and K are the arc centerpoint coordinates, F sets the Z-axis feedrate.

The program to mill the part in Figure 12-7 is presented in Figure 12-9.

Program Explanation

N001

Safety line.

N101

G00X1.6Y0. – Positions spindle to the start point.
S800M03 – Turns on spindle clockwise at 800 rpm.

N102

G45Z-.818H01 – Rapids spindle to starting depth and turns on the tool length offset.
M08 – Turns on the coolant.

N103

G01X.9694F7. – Feeds the cutter into the workpiece at a feedrate of 7.0 inches per minute.

```
%
O1209
N001G00G80G90G40                   (SAFETY LINE)
N101G00X1.6Y0.S800M03              (POSITION TO START POINT)
N102G45Z-.818H01M08               (FEED TO START DEPTH)
N103G01X.9694F7.                  (FEED INTO WORKPIECE)
N104G17G02X.9694Y0.Z-.868I-1.J0.  (FIRST HELICAL TURN)
N105G02X.9694Y0.Z-.918I-1.J0.     (SECOND HELICAL TURN)
N106G02X.9694Y0.Z-.968I-1.J0.     (THIRD HELICAL TURN)
N107G01X1.6                       (FEED OUT OF WORKPIECE)
N108G00G91G28Z0.M09               (Z-AXIS TO HOME ZERO)
N109G28X0.Y0.M05                  (X/Y AXES TO HOME ZERO)
N110M30                           (END PROGRAM)
%
```

FIGURE 12-9
Program to mill part in Figure 12-7

N104

G17 – Specifies the X/Y plane as the circular interpolation plane.

G02 – Initiates circular interpolation clockwise.

X.9694Y0. – The end coordinates of the circular cut (in this case 360 degrees).

Z-.918 – The Z coordinate for the helical cut. The Z axis will feed downward to this coordinate during the circular X/Y motion.

I-1.J0. – The X/Y vector from the center of the cutter to the center of the arc.

N105

G02 – Circular interpolation code.

X.9694Y0. – The end coordinates of the circular cut (in this case 360 degrees).

Z-.918 – The Z coordinate for the helical cut.

I-1.J0. – The X/Y vector from the center of the cutter to the center of the arc.

N106

G02 – Circular interpolation code.

X.9694Y0. – The end coordinates of the circular cut (in this case 360 degrees)

Z-.968 – The Z coordinate for the helical cut.

I-1.J0. – The X/Y vector from the center of the cutter to the center of the arc.

N107

G01 – Puts the control back in linear interpolation mode.

X1.6 – Feeds the spindle clear of the part.

N108

G00G91G28Z0.M09 – Sends the spindle to Z home position and turns off the coolant.

N109

G28X0.Y0.M05 – Sends the spindle to the X/Y home position and turns off the spindle.

N110

M30 – End of program code.

SUMMARY

The important concepts presented in this chapter are:

- Mirror imaging means changing the sign (+ or −) of an axis movement.
- Mirror imaging is used in a program to save repetitive programming when the direction of movement is the only difference between part features.
- Mirror imaging is normally used in conjunction with subroutines or do loops.
- Polar rotation is an indexing of the NC machine's Cartesian coordinate system to some angle other than its normal state.
- Polar rotation may be used to perform operations that otherwise would require the use of a rotary axis or lengthy coordinate calculations.
- Polar rotations may be used in conjunction with do loops or subroutines.
- Helical interpolation is circular interpolation with two axes while simultaneously feeding at a linear rate with the third. The result of this type of operation is a helix.
- Care must be taken in calculating the number of turns and the lead of a helix, be it a thread or other type of part.
- Helical interpolation may be used inside of or in conjunction with do loops and subroutines.

VOCABULARY INTRODUCED IN THIS CHAPTER

Helical interpolation
Mirror imaging
Polar axis system
Polar rotation
Thread lead

REVIEW QUESTIONS

1. What is mirror imaging? Why is it used?
2. When would mirror imaging be used in a program?

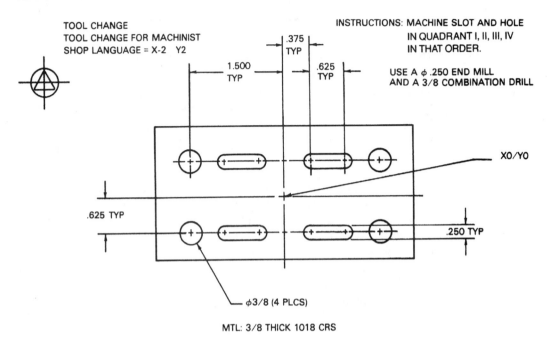

TOOL CHANGE
TOOL CHANGE FOR MACHINIST
SHOP LANGUAGE = X-2   Y2

.375 TYP

1.500 TYP

.625 TYP

INSTRUCTIONS: MACHINE SLOT AND HOLE
IN QUADRANT I, II, III, IV
IN THAT ORDER.

USE A φ .250 END MILL
AND A 3/8 COMBINATION DRILL

X0/Y0

.625 TYP

.250 TYP

φ3/8 (4 PLCS)

MTL: 3/8 THICK 1018 CRS

FIGURE 12-10
Part drawing for review question #3

3. Write a program to mill the slots and drill the holes in the part shown
   in Figure 12-10.
4. What does polar rotation do?
5. What types of equipment can polar rotations substitute for?
6. Can polar rotations be used with subroutines and do loops?
7. What type of information must be given in the program for the MCU
   to perform polar rotation?
8. Write a program to mill the slots in the part shown in Figure 12-11.
9. What is helical interpolation?
10. When would helical interpolation be useful in a program?

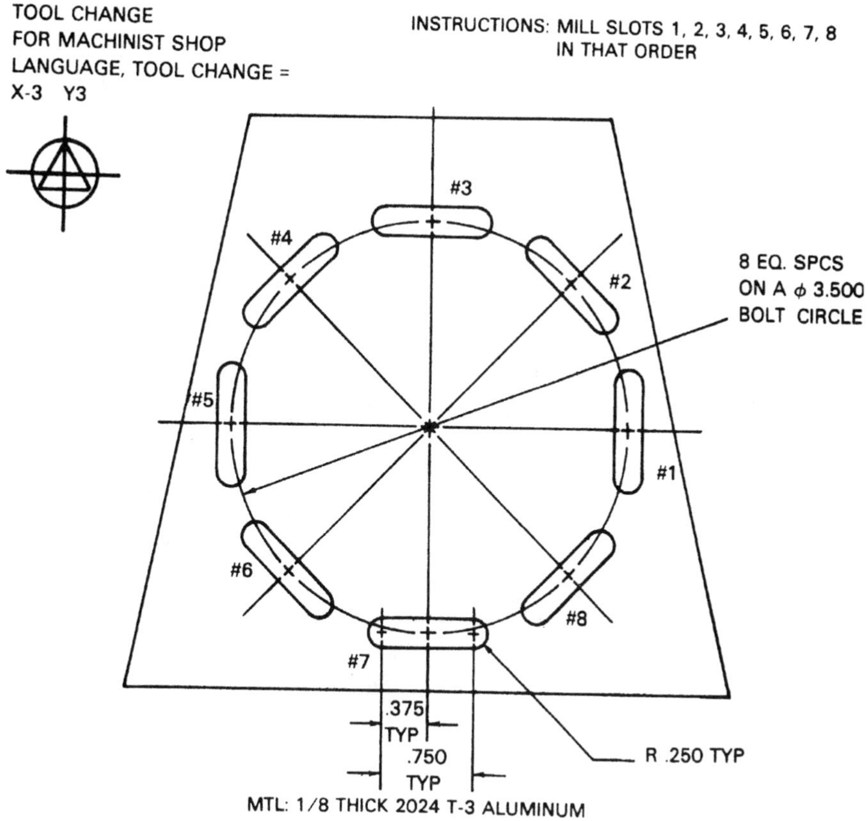

TOOL CHANGE
FOR MACHINIST SHOP
LANGUAGE, TOOL CHANGE =
X-3 Y3

INSTRUCTIONS: MILL SLOTS 1, 2, 3, 4, 5, 6, 7, 8
IN THAT ORDER

#3

#4

#2

8 EQ. SPCS
ON A φ 3.500
BOLT CIRCLE

#5

#1

#6

#8

#7

.375
TYP
.750
TYP

R .250 TYP

MTL: 1/8 THICK 2024 T-3 ALUMINUM

FIGURE 12-11
Part drawing for review question #8

CHAPTER 13

The Numerical Control Lathe

OBJECTIVES Upon completion of this chapter, you will be able to:

- Describe the difference between a conventional lathe bed arrangement and a slant bed arrangement, listing the advantages of the slant bed for NC.
- Explain axis movement on a CNC lathe.
- Describe the method of toolholding used on CNC turning machines.
- Explain what a tool offset number is.
- Describe two methods of tool selection used on CNC turning machines.
- Describe how spindle speed is designated on gear head and variable speed lathes.
- Explain how feedrates are specified on CNC turning equipment.
- Define TNR.

Up to this point, the programming features of CNC mills have been discussed, but numerical control is used for turning equipment as well. In the milling examples, both Machinist Shop Language and word address formats were given. For the turning programs discussed in Chapter 14, only word address format will be used. The coding will be a version used with FANUC lathe controllers, designed to be generic and so to illustrate the basic programming steps involved. A numerical control lab in a school will have equipment that differs in one way or another from that presented here. Students are advised to familiarize themselves with the codes used for the machines they will be using.

LATHE BED DESIGN

Older NC lathes, and those that have been converted to numerical control with retrofit units, look like traditional engine lathes. The lathe carriage rests on the ways. The ways are in the same plane and are parallel to the floor, as illustrated in Figure 13-1. This arrangement allows the machinist to reach all the controls readily. Since the CNC lathe performs its operations automatically, this type of arrangement is not necessary. In fact; it is quite

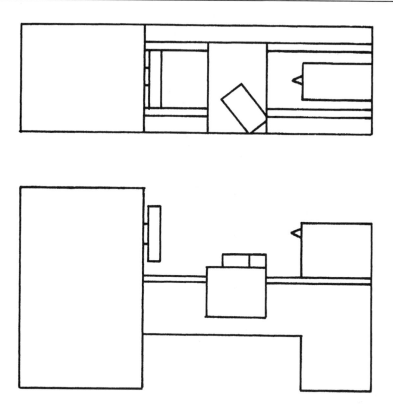

FIGURE 13-1
Bed arrangement on a conventional lathe

awkward, as the operator will be busy with other responsibilities while the program is running and so will not necessarily be there to brush the chips off the ways. In a conventional lathe bed arrangement, the chips have nowhere to fall except on the ways. To overcome this problem, many CNC lathes make use of the slant bed design illustrated in Figure 13-2.

On many NC lathes, the turret tool post is mounted on the opposite side of the saddle, compared to a conventional lathe, to take advantage of the slant bed design. The slant bed allows the chips to fall into the chip pan, rather than on tools or bedways. Despite its odd appearance, the slant bed NC lathe functions just like a conventional lathe. Figures 13-3 and 13-4 show modern CNC turning machines. Notice the slant bed arrangement.

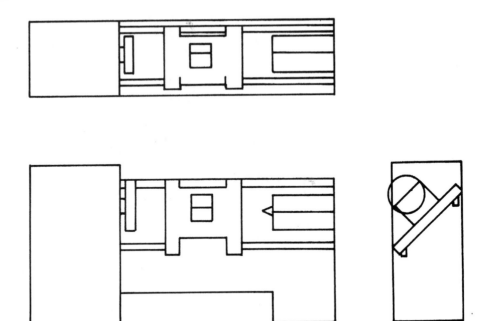

FIGURE 13-2
Slant bed for NC or CNC lathe

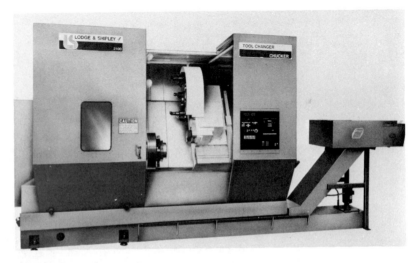

FIGURE 13-3
A modern CNC turning center employing automatic tool change *(Photo courtesy of Lodge and Shipley Co.)*

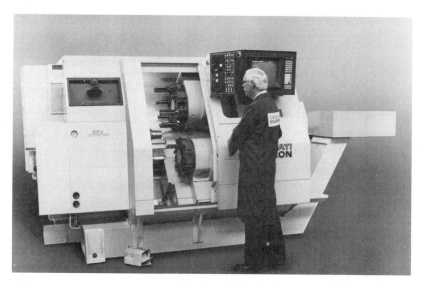

FIGURE 13-4
A four-axis CNC turning center *(Photo courtesy of Cincinnati Milacron)*

AXIS MOVEMENT

The axis movement of a basic CNC lathe is diagrammed in Figure 13-5. Some turning machines, such as that shown in Figure 13-4, are four-axis machines. In this book, only the basic two-axis machine is programmed. The programming concepts learned on a two-axis machine are the foundation necessary to program more complex machinery.

The basic lathe has only two axes, X and Z. Since the Z axis is always parallel to the spindle, longitudinal (carriage) travel is designated Z. The cross slide movement is designated X, since it is the primary axis perpendicular to Z. If it were possible to move the carriage up and down, that axis would be Y. There is, however, a potential problem with this arrangement. There appear to be two Z axes: the carriage movement and the tailstock movement. To eliminate this problem the tailstock is usually called the W axis on lathes with programmable tailstocks. Programmable tailstocks, which are rear turret assemblies on CNC equipment, are the third and sometimes fourth axis on more complex equipment. The turning center in Figure 13-4 has two programmable saddles. In such cases the axes of the second saddle are usually designated

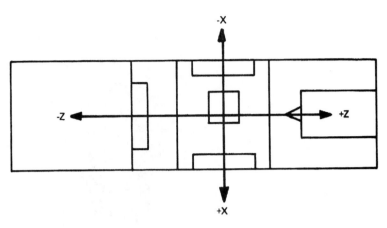

FIGURE 13-5
Lathe axis movement

W and U, with W being saddle travel and U being cross slide travel. There are some imported lathes on which the X-axis direction is reversed. The programmer must determine if such a situation exists before writing the lathe program.

TOOLHOLDERS AND TOOL CHANGING

Either a rigid toolholder or a tool turret is used to hold the tools on an NC lathe. Figure 13-3 shows a CNC chucker employing a rigid toolholder. The turning center in Figure 13-4 employs a tool turret, in which the various tools needed for lathe operations are placed in toolholders. When a tool change is necessary, the appropriate turret is indexed to the next tool needed. Simple lathes use six-sided turrets; larger turning machines use eight-, ten-, and twelve-sided turrets.

With the development of robotics, new tool changing and work handling schemes are appearing. Figure 13-6 shows a robot arm used for handling workpieces, and Figure 13-7 illustrates the robot in operation. To teach the basics of CNC programming, this text will focus on nonrobotic tool change.

The toolholders used on NC turning machines are of very rigid design. The tools used for turning are of the carbide insert type, made to much more exacting tolerances than conventional lathe insert tooling.

A tool change command in a turning program either changes the turret position or causes an automatic tool change, depending on the type of machine used.

FIGURE 13-6
A robot arm used for part load and unload *(Photo courtesy of Cincinnati Milacron)*

FIGURE 13-7
A robot arm in action *(Photo courtesy of Cincinnati Milacron)*

Automatic Tool Change

In a CNC turning program for a machine with a rigid toolholder, M06 is used to initiate an automatic tool change. The T address is used (as it is in milling programs) to specify the desired tool. The T address also calls up the tool offsets. The format for automatic tool change is:

M06 T n1 n2

Where M06 initiates the tool change, T is the tool address, n1 is the tool number, and n2 is the tool offset number.

Turret Position

T is used in a similar manner with turret tool selection. The format is:

Tn1 n2

Where the first number is the turret position and the second is the tool offset number.

Since one tool may be used in several positions, a turret position is used rather than a tool number. The turret position corresponds to the turret station number. T01 will index the tool in station one into position. Some NC lathes can utilize more than one tool on a single station. It is possible, therefore, for T0101 to refer to one tool and T0111 to refer to another. This is referred to as piggybacking a tool station.

One other point should be kept in mind when changing tools: the carriage (or tailstock) does not necessarily move to a tool change location. It is often necessary, therefore, first to move the carriage or tailstock turret out of the way before making a tool change. It may also be necessary to program a dwell (G04) to halt the program, giving the tool time to index to position safely.

Tool Offset Numbers.   Each turning tool used on a lathe has a radius. When programming the coordinates for a location, the centerline of the tool radius is programmed. Tool offsets allow the center of the tool nose to be programmed and thus compensate for minor differences in length that exist between tools, and the effects of tool wear.

When the tools are set up, the operator enters the offsets into the tool registers. When the offsets are active, the MCU compensates for them, eliminating the need for premeasured tooling. In this manner, the programmer can treat all tools as being the same length, just as in milling.

The *offset number* is the number of the register in which a particular tool's offset is stored. Generally the register number will match the tool number.

The tool information entered manually prior to the start of the program is entered in this form:

X Offset
  Z Offset
  Tool Nose Radius
  Standard Tool Nose Vector Number

Tool Nose Radius and Standard Tool Nose Vector Numbers

The tool nose radius and tool nose vector numbers are optional. They are entered if using cutter comp. Cutter diameter compensation is called *tool nose radius compensation* (TNR comp) on turning machines. The tool radius tells the MCU the amount of compensation that is to be used. With NC machining centers this value was entered in a comp register.

TNR comp is utilized just as cutter comp was in Chapter 10. It can be used to program the part line, or fine tune the tool path to compensate for tool wear. The major difference is lathe tools are not completely circular as is a milling cutter. To aid in proper compensation of the tool path and correctly identify alarm conditions, a tool nose vector number is entered in the register. Tool nose vector numbers tell the MCU the orientation of the tool nose. Figure 13-8 shows the various directions in which a tool may be oriented. These directions are referred to as vectors. Each vector has a number associated with it that is used to describe the tool orientation to the MCU.

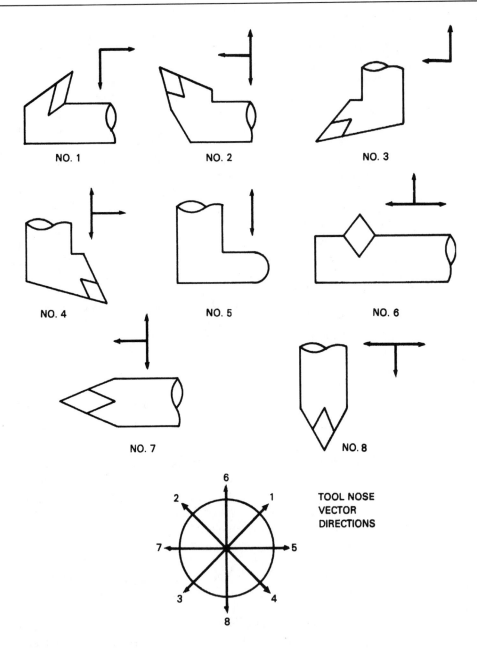

NO. 1   NO. 2   NO. 3

NO. 4   NO. 5   NO. 6

NO. 7   NO. 8

6
2   1
7   5
3   4
8

TOOL NOSE
VECTOR
DIRECTIONS

FIGURE 13-8
Tool nose vectors

Tool Edge Vs Centerline Programming

The tool nose may be programmed in one of two ways when TNR comp is not active: by the tool edge or by the tool nose radius centerline.

Tool edge programming is adequate for simple straight line cuts where the part surfaces intersect each other at right angles. Problems are encountered, however, when angles and especially arcs are programmed this way. Figure 13-9A illustrates this point. If the tool edge is programmed, the I and K centerpoints of the illustrated arc must be shifted. This results in a tool path that does not follow the desired arc exactly. The amount of error that is induced depends upon the size of the cutter and the radius of the arc. In any case, tool edge programming should not be used when encountering arcs and angles.

Tool centerline programming is identical to the centerline programming done when milling. Figure 13-9B demonstrates how the cutter centerlines and part surface centerlines coincide when the center of the tool nose radius is programmed. This type of programming is demonstrated in Chapter 14.

SPINDLE SPEEDS

Spindle speed is specified using an S address, just as in milling. On turning machines with a gear head design, the spindle speed is changed by shifting gears in the headstock. On gear head machinery, there are usually two or more gear ranges. An M function is used to select the gear range in which the desired speed is located. M40 through M46 generally serve this purpose. For gear head examples in this text, M40 will be used for low range, M41 for mid range, and M42 for high range.

The following chart shows a sample of speed ranges for gear head machines. This chart is not for a particular machine but is representative of the type of spindle speed spread found on a machine.

LOW RANGE

10	15	20	25
30	40	50	65
75	90	110	125

MEDIUM RANGE

55	70	95	120
140	155	175	200
235	260	290	300

HIGH RANGE

285	335	380	450
530	660	900	1200
1800	2100	2500	3000

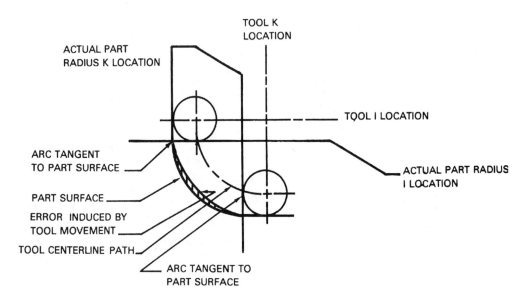

ACTUAL PART
RADIUS K LOCATION

TOOL K
LOCATION

TOOL I LOCATION

ARC TANGENT
TO PART SURFACE

ACTUAL PART RADIUS
I LOCATION

PART SURFACE

ERROR INDUCED BY
TOOL MOVEMENT

TOOL CENTERLINE PATH

ARC TANGENT TO
PART SURFACE

FIGURE 13-9 (A)
Error induced by programming tool edge

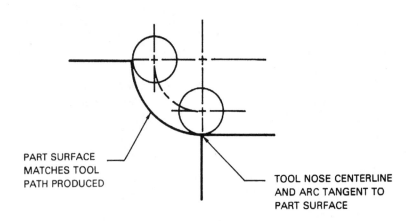

PART SURFACE
MATCHES TOOL
PATH PRODUCED

TOOL NOSE CENTERLINE
AND ARC TANGENT TO
PART SURFACE

FIGURE 13-9 (B)
Tool nose centerline programming

Some CNC turning machines use a variable speed drive with which an infinite number of speeds are available between the highest and lowest speeds. In these cases the speed is selected using the S address as it is in milling.

FEEDRATES

With a CNC lathe, assigning feedrates is quite simple. A G98 or G94 code (depending on the controller) tells the MCU that the following feedrate is in inches per minute. For example, G98 F7 specifies a feedrate of 7 inches per minute. A G99 or G95 in a turning program specifies a feedrate in inches per revolution. For example, G99 F.015 specifies a feedrate of .015 inch per revolution. Appendix 3 contains a list of common G codes used in lathe programming.

MACHINE ORIGIN AND WORK COORDINATE SYSTEMS

An NC lathe generally has a fixed-zero position assumed by the executive program upon power-up. This position is known as the home zero or machine origin. The physical location of this position varies from controller to controller and machine model to machine model. It is usually one of two locations: X0 = centerline of the spindle, Z0 = the chuck mounting surface of the spindle, or X0 = extreme X + location, Z0 = extreme Z + location. It is usually necessary to establish a zero point on the part different from the machine origin location. This position is called the *work coordinate system* or *part zero.* There are two methods used to accomplish this.

The first method involves the use of an axis preset command—G50. The G50 transfers the zero point from the home zero to the coordinates specified with the command. The format for a G50 command is:

G50 Xxx.xxxx Zzz.zzzz

Where: G50 = the axis preset command
           xx.xxxx = the X-axis distance to the part zero
           zz.zzzz = the Z-axis distance to the part zero

A G50 command is issued at the start of each tool. Since the programmer will not know the axis preset distances in advance, zeros should be used or some other prearranged value in the G50 line. The actual values will be determined by the setup person and edited in the control when the job is set up.

The second method uses registers called *work coordinates.* These are registers in the MCU that tell the MCU the distance from home zero to the part zero. If a machine has more than one available work coordinate, multiple zero points may be used for complex programming. Another advantage to multiple work coordinates is the ability to have more than one program loaded in the MCU, each with its own work coordinate. This is a decided advantage when running several repeating jobs through a turning center. Each work coordinate is called by a G code. If a program were to use four work coordinates, they would be selected by the codes G54, G55, G56, and G57. The first work coordinate (G54 in this case) is the default work coordinate. This work coordinate is automatically activated upon power-up. If using only the default work coordinate, the G code may be omitted. The work coordinate values are entered by the setup person when the job is prepared. The programmer must instruct the setup personnel the position on the part of the part zero location.

QUICKSETTERS

A fairly recent development has been the use of quicksetters—arms with tool sensors on them. During job setup the arm is lowered into position, the operator jogs a tool to the presetting position and touches it off on the sensor. The quicksetter automatically sets the values of the work coordinate and the tool offset registers.

SUMMARY

The important concepts presented in this chapter are:

- CNC turning machines often use a slant bed arrangement to protect the machine ways from chips. Although different in appearance, the functioning of a slant bed and conventional bed machine is identical.
- There are two basic axes, X and Z, on a CNC lathe. If the lathe has additional axes, they are generally designated U and W.
- TNR stands for tool nose radius compensation. TNR is the equivalent in CNC turning to cutter diameter compensation in milling.
- A tool turret or a rigid toolholder is used to hold the tools on an NC lathe.
- Tool offsets are entered into the MCU prior to running the program to compensate for minor setup adjustments.
- A standard tool nose vector number is used to identify the orientation of a particular tool when using TNR.
- A tool change command in turning programs will either change the turret position or cause an automatic tool change, depending on the type of machine used.
- The tool change format for turret changing is: T n1 n2
  Where T is the tool change command, n1 is the turret position and n2 is the tool offset number.

- The format for automatic tool change is: M06 T n1 n2
Where M06 initiates the tool change, T is the tool address, n1 is the tool number, and n2 is the tool offset number.
- Spindle speeds are specified directly using the S address. On gear head machines, it is necessary to specify the gear range when selecting a range outside the active one.
- Feedrates on CNC lathes can be specified either in inches per minute (using G94 or G98), or in inches per revolution (using G95 or G99).
- To set a part at X0/Z0 point, it is necessary to transfer the machine origin to the workpiece using a G code.

VOCABULARY INTRODUCED IN THIS CHAPTER

Centerline programming
Lathe bed
Quicksetter
Slant bed
Tool edge programming
Tool nose radius
Tool nose vector numbers
Tool offset numbers
Tool turret
Turret position

REVIEW QUESTIONS

1. What is the difference between a slant and conventional lathe bed arrangement? What is the advantage of a CNC slant bed lathe?
2. Draw a sketch illustrating the axis movement on a lathe.
3. What type of toolholding is used on CNC turning machines?
4. What is the purpose of a tool offset register?
5. What is a standard tool nose vector number?
6. What types of turrets in addition to the four-sided turret are used on CNC lathes?
7. How are spindle speeds designated on CNC turning machines? What additional coding is required on gear head machines?
8. How are feedrates specified on CNC lathes? What codes are used?
9. What is the format for a turret tool change command? For an automatic tool change?
10. What does TNR stand for?
11. What is the machine origin?
12. How is an X0/Y0 point established on a workpiece?

CHAPTER 14

Programming CNC Turning Machines

OBJECTIVES Upon completion of this chapter, you will be able to:

- Write simple turning and facing routines.
- Write simple taper turning routines.
- Write simple routines to perform circular interpolation, using programmed arc centers and programmed radius value methods.
- Write simple thread-turning routines using single pass and multipass threading.

CNC lathe controllers vary in their coding to an even greater extent than mill controllers. It is, therefore, difficult to discuss programming practices. EIA standards specify axis movement, for example, but some lathes use a left-hand coordinate system, with the X and Z axes reversed from the standard configuration. Other lathes reverse the X axis direction and not the Z. On lathes using twin turrets, the X axis is often reversed. The uses of coding and the cycles available also differ to a large extent. The EIA codes pertaining to lathes are generally used, but many other codes may be added.

This chapter will discuss basic lathe programming routines for turning, facing, taper turning, circular interpolation, and thread cutting. Each routine is placed in a miniprogram. Each program can be thought of as a building block; to machine a complete part, these building blocks can be linked together in one program as will be demonstrated.

MACHINE REFERENCE POINT

A machine *reference point* is a fixed position on the machine. Upon receiving the proper G code, the machine automatically returns to the reference point location. This point is often the home zero location, used for tool changing and as a park position at the end of the program. Often it is necessary to send the tool back to the reference point by way of another point, called an *intermediate point.* The code used in this chapter to return the tool to reference is G28. A U and W coordinate is specified along with the G28. Upon receiving the G28, the tool moves to the reference point.

DIAMETER VS RADIUS PROGRAMMING

The difference between radius programming and diameter programming is an important one. *Diameter programming* references the X-axis coordinate to the diameter of the workpiece. This means that every .001 inch programmed moves the tool .0005 inch as measured radially. If the X axis advances .500 inch into the part, .500 inch is removed from the diameter. To accomplish this, the X axis moves only .250 inch, or half the programmed amount.

In *radius programming,* the X axis moves the programmed amount. If .500 inch of movement along the X axis is programmed, the tool advances .500 inch. When the Z-axis move is made, 1.000 inch of material is removed from the part.

Canned cycles on a machine call for the information to be entered in either diameter or radius coordinates, depending on the cycle's function. The machine manual must be consulted to determine the type of coordinate expected. The coordinates may be either incremental or absolute, depending on whether G90 or G91 is active. As in milling, G90 selects absolute positioning and G91 selects incremental. Other controllers use a "W" address for incremental X and a "U" address for incremental Z.

TURNING AND FACING

Figure 14-1 shows a part to be turned and faced in a lathe. Note that the position of the tool turret relative to the X0/Y0 location and the machine origin is given. The machine coordinate system may be transferred to the workpiece either within the program by use of G codes or by the operator during machine setup. It is usually more efficient to define the work coordinate system during setup. For routines in this chapter, this will be assumed. Figure 14-2 shows a part similar to the one in Figure 14-1 but with metric dimensions. Figure 14-3 presents a short program to turn and face the part drawn in Figure 14-1. Figure 14-4 presents a metric version. To program this part, the following codes will be used (see page 288):

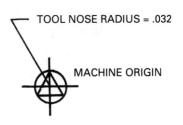

MACHINE ORIGIN

TOOL NOSE RADIUS = .032

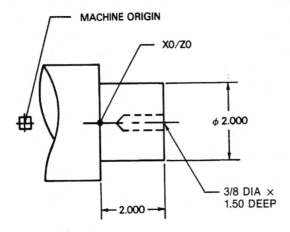

MACHINE ORIGIN

X0/Z0

ϕ 2.000

3/8 DIA $\times$
1.50 DEEP

2.000

MTL: ϕ 2.5   CRS

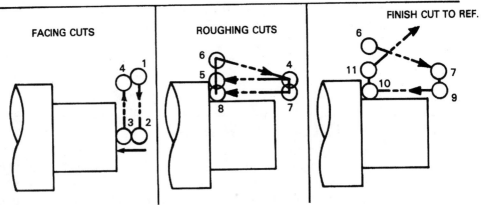

FACING CUTS

ROUGHING CUTS

FINISH CUT TO REF.

FIGURE 14-1
Part to be turned and faced in a lathe

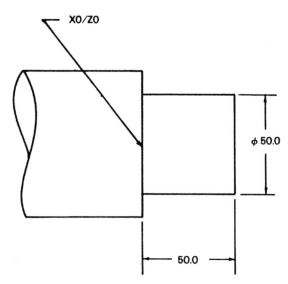

FIGURE 14-2
Part from Figure 14-1 with metric dimensions

```
%
O1403
(* **********)
(* X0 = CENTERLINE OF SPINDLE)
(* Z0 = PART SHOULDER)
(* **********)
N010G00G99M08              (SAFETY LINE, COOLNT ON)
N020T0101M42               (TURRET POS, HIGH RANGE)
N030S1200M03               (SPINDLE ON)
N040X2.6Z2.042             (POSITION TO #1)
N050G01X0.F.007            (FEED TO #2)
N060Z.032                  (FEED TO #3)
N070X2.314F.003            (FEED TO #4)
N080Z.042F.007             (FEED TO #5)
N090X2.6                   (FEED TO #6)
N100G00X2.320 Z2.132       (RAPID TO #4)
N110G01X2.0840             (FEED TO #7)
N120Z.042F.003             (FEED TO #8)
N130X2.6                   (FEED TO #6)
N140G00X2.084 Z2.132       (RAPID TO #7)
N150G00X2.062              (FEED TO #9)
N160Z.032F.003             (FEED TO #10)
N170X2.55                  (FEED TO #11)
N180G00G28U0.W0.M09        (RETURN TO HOME/COOLNT OFF)
N190M05                    (SPINDLE OFF)
N200M30                    (END PRGM)
%
```

FIGURE 14-3
Program to turn and face part in Figure 14-1

```
%
O1404
(* **********)
(* X0 = CENTERLINE OF SPNIDLE)
(* Z0 = PART SHOULDER)
(* **********)
N010G00G99M08              (SAFETY LINE)
N020T0101M42               (TURRET POS, HIGH RANGE)
N030S1200M03               (SET SPEED)
N040X67.Z52.               (POSITION TO #1)
N050G01X0.F.5              (FEED TO #2)
N060Z51.                   (FEED TO #3)
N070X60.F.13               (FEED TO #4)
N080Z2.F.5                 (FEED TO #5)
N090X67.                   (FEED TO # 6)
N100G00X60.Z101            (RAPID TO #4)
N110G01X53.                (FEED TO #7)
N120Z2 F.13                (FEED TO #8)
N130X67.                   (FEED TO #6)
N140G00X53.Z101.           (RAPID TO #7)
N150G01X51.                (FEED TO #9)
N160Z1.                    (FEED TO #10)
N170X66.                   (FEED TO #11)
N180G00G28 U0.W0.M09       (RETURN TO REF & COOLNT OFF)
N190M05                    (SPINDLE OFF)
N200M30                    (END PGRM)
%
```

FIGURE 14-4
Program to turn and face part in Figure 14-2

G00—As in milling programs, G00 puts the machine in rapid traverse mode.

G01—Linear interpolation. As with milling, the machine will position the tool to the programmed coordinates at feedrate, in a straight line.

G28—Return to reference point. A G28 is programmed with a U and W coordinate. Upon receiving the G28, the machine positions the tool at the fixed machine reference point.

G99—Selects inches per revolution or millimeters per revolution feedrates. The feedrates are the programmed value per revolution of the spindle. A G95 F.01 advances the tool .010 inch for every revolution of the spindle.

M40—Selects the low gear range.

M41—Selects the middle gear range.

M42—Selects the high gear range.

Program Explanation (Refer to Figures 14-3 and 14-4.)

A .032-inch tool nose radius is used on the tool in the nonmetric program. A 1-mm tool nose radius is used in the metric program. One roughing and one finish facing cut are used; two roughing and one finish turning cuts are used.

N010

G00 – Selects the rapid traverse mode.

G99 – Selects per revolution feedrate.

M08 – Turns on the coolant.

N020

T0101 – Selects a tool number and calls the tool offset in register # 1.

M42 – Selects high gear range.

N030

S1200 – Sets the spindle speed to 1200 RPM.

M03 – Turns on the spindle.

N040

X/Z coordinates – Rapid the tool to location #1, Figure 14-1. The X axis coordinate is diameter programmed, as are all the X coordinates in this program.

N050

G01 – Selects feedrate movement.

X0 – Feeds the tool to location #2. This is the rough facing cut.

F.007 – Sets the feedrate to .007 inch per spindle revolution (.5 mm metric.)

N060

Z coordinate – Feeds the tool from location #2 to location #3. This sets the Z axis depth for the finish facing cut.

N070

X coordinate – Feeds the tool from location #3 to location #4. The coordinate is diameter programmed.

F.003 (F.13 metric) – Sets finish feedrate.

N080

Z coordinate – Feeds the tool from location #4 to location #5. This is the first roughing pass.

F.007 (F0.5 metric) – Sets the roughing pass feedrate.

N090

X coordinate – To feed from location #5 to location #6. This cut rough faces the shoulder of the part and retracts the tool for the return move.

G00 – Selects rapid traverse. This is a return to start of cut move. No feedrate is necessary.

X/Z coordinates – Move the tool at rapid from location #6 to location #4.

N110

G01 – Selects linear interpolation (feedrate mode).

X coordinate – Feeds the tool from location #4 to location #7. This move could also have been made in rapid traverse. Using a feedrate here eliminated the possibility of chipping the tool cutting edge on the corner of the stock.

Z coordinate – Feeds the tool from location #7 to location #8. This is the second rough turning pass.

F.003 (F.13 metric) – Sets finish feedrate.

N130

X coordinate – Rough faces the shoulder, retracting the tool.

N140

G00 – Selects rapid traverse.

X/Z coordinate – Positions the tool to location #7.

N150

G01 – Selects feedrate movement.

X coordinate – Feeds the tool from location #7 to location #9. This positions the X axis depth for the finish pass.

N160

Z coordinate – Feeds the tool from location #9 to location #10. This completes the turning.

N170

X coordinate – Feeds the tool from location #10 to location #11. This move finish faces the part shoulder.

N180

G00 – Selects rapid traverse.

G28U0.W0. – Initiates a return to reference.

M09 – Turns the coolant off.

N190

M05 – Turns the spindle off.

N200

M30 – Signals the end of program.

TAPER TURNING

Linear interpolation on a lathe is used to turn tapers. It is similar in use to linear interpolation to cut angles when milling. On the part pictured in Figure 14-5 is a taper to be bored. The part is a steel casting, requiring that the taper be rough and then finish machined. (The short program to perform these operations is shown in Figure 14-7.)

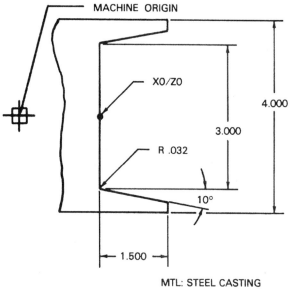

MACHINE ORIGIN

X0/Z0

R .032

4.000

3.000

10°

1.500

MTL: STEEL CASTING

TOOL RADIUS = .032

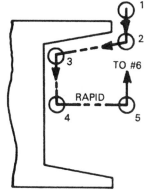

1

2

3

TO #6

RAPID

4   5

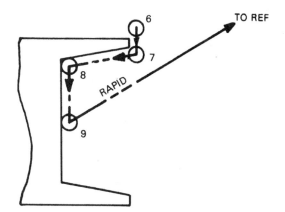

6

TO REF

7

8

RAPID

9

FIGURE 14-5
Taper turning

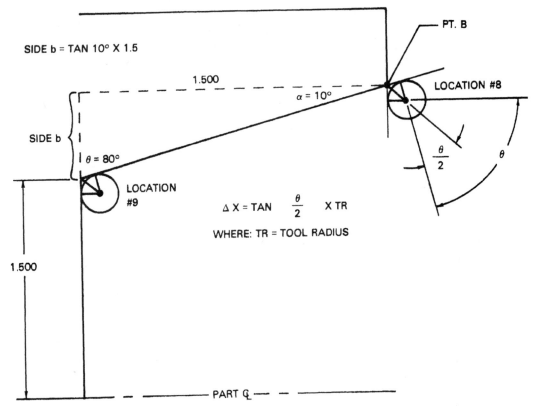

SIDE b = TAN 10° X 1.5

1.500

$\alpha = 10°$

PT. B

LOCATION #8

SIDE b

$\theta = 80°$

LOCATION #9

$\frac{\theta}{2}$

θ

$\Delta X = TAN \quad \frac{\theta}{2} \quad X\ TR$

WHERE: TR = TOOL RADIUS

1.500

PART ℄

FIGURE 14-6
Determining cutter offsets

Cutter offset calculations necessary with taper turning are similar to those used when calculating angle cuts for milling. Figure 14-6 depicts the relationship of the lathe tool nose to the tapered part surfaces. Two coordinate locations require cutter offsets. Both locations present the identical situation, so that calculating one offset will automatically yield the other. This is the same simple cutter-to-angle relationship first discussed in Chapters 8 and 9, and the formula given in Appendix 6, Figure 1, can be used. In this case, the Y axis in the formula is the X axis on the lathe, and the X axis in the formula is the Z axis on the lathe. The offset is calculated as follows, where CR is the tool nose radius:

$$X = TAN \left(\frac{\theta}{2}\right) \times CR$$

$$X = TAN\ 40 \times .032$$

$$X = .8391 \times .032$$

$$X = .02685\ or\ .027$$

Before the cutter offset can be used, however, it is necessary to calculate the location of point B, Figure 14-6. By solving the indicated triangle for side b and adding that length to the known radius of the taper (1.5 inches), the radius dimension from the part center line to point B can be determined.

$$\frac{b}{1.5} = \text{TAN } 10$$

$$b = \text{TAN } 10 \times 1.5$$

$$b = .1763 \times 1.5$$

$$b = .26445 \text{ or } .264$$

The value of .264 added to the 1.5 radius gives a distance of 1.764 from the part centerline to point B. The cutter offset can be subtracted from the 1.764 distance to find the dimension from the part centerline to cutter location #7. This distance is 1.737. The X coordinate for this location, however, will be diameter programmed. The 1.737 must now be doubled to arrive at the X coordinate to be programmed, or 3.474.

The calculated tool offset can also be subtracted from the 1.5 known radius to arrive at the 1.473 dimension from the part centerline to tool location #8. Doubling this distance gives 2.946, the X axis coordinate for location #8. The offset for the Z axis in both these cases is simply the radius of the tool nose.

```
%
O1407
(* **********)
(* X0 = CENTERLINE OF SPINDLE)
(* Z0 = PART SHOULDER )
(* **********)
N010G00G99M08                   (SAFETY LINE, COOLNT ON)
N020T0101M42                    (TURRET POS, HIGH RANGE)
N030S800M03                     (SPINDLE ON)
N040X4.1Z1.51                   (POSITION TO #1)
N050G01X3.454F.007              (FEED TO #2)
N060X2.974Z.042                 (FEED TO #3)
N070X0.                         (FEED TO #4)
N080G00Z1.542                   (RAPID TO #5)
N090X4.1 Z1.532                 (RAPID TO #6)
N100G01X3.474F.003              (FEED TO #7)
N110X2.946Z.032                 (FEED TO #8)
N120X0.                         (FEED TO #9)
N130G00U0.W0.M09                (RETURN TO REF & COOLNT OFF)
N140M05                         (SPINDLE OFF)
N150M30                         (END PRGM)
%
```

FIGURE 14-7
Program to turn part in Figure 14-5

Program Explanation (Refer to Figure 14-7)

N010

G00 – Selects rapid traverse.

G99 – Specifies inches per revolution feedrate.

M08 – Turns the coolant on.

N020

T0101 – Select the tool and the offset.

M42 – Selects high gear range.

N030

S800 – Sets the spindle speed to 800 RPM.

M03 – Turns the spindle on.

N040

X4. 1 Z1.51 – Position the tool to location # 1, Figure 14-5.

N050

G01 – Selects linear interpolation. The tool will feed in a straight line between the next coordinate programmed and the current tool location.

X3.454 – Feeds the tool from location #1 to location #2. This coordinate was determined by adding approximately the desired amount of finished stock to the cutter coordinate of location #8, calculated previously.

F.007 – Sets the feedrate.

N060

X2.974 Z.042 – Coordinates to feed the tool from location #2 to location #3. The X coordinate was determined by subtracting .020 from the calculated finished location coordinate. Although this coordinate will not leave exactly .010 inch of stock per side to be removed during finishing, the amount left will be close to that.

N070

X0. – Feeds the tool from location #3 to location #4.

N080

G00 – Selects rapid traverse.

Z1.542 – Sends the tool at rapid to location #5. This is an intermediate location used before sending the tool to location #6. If the tool were moved from location #4 to location #6 directly, the corner of the part would be cut off. Laying a straightedge between location #4 and location #6 will demonstrate the point.

N090

X4.1 Z1.532 – Feeds the tool from location #5 to location #6 at rapid (G00 is active).

N100

G01 – Selects linear interpolation.

X3.474 – Feeds the tool from location #6 to location #7. This is the co-ordinate location calculated earlier.

F.003 – Sets the finish pass feedrate to .003 inch per revolution.

N110

X2.946 Z.032 – Coordinates of location #8.

N120

X0 – Feeds the tool from location #8 to location #9.

N130

G00 – Specifies rapid traverse.

G28U0.W0. – Initiates a return to reference.

M09 – Turns the coolant off.

N140

M05 – Turns the spindle off.

N150

M30 – Ends the program.

CIRCULAR INTERPOLATION

Circular interpolation on a lathe does not differ significantly from circular interpolation when milling. There are two ways that an arc center can be programmed using CNC turning machines. The centerpoint can be programmed using I and K, or the center may be specified on some machinery as a radius value. Some machining centers may have an arc centerpoint specified by the radius method also. When I and K are used, I is programmed as the X-axis coordinate of the arc centerpoint, and K is programmed as the Z-axis coordinate. The format is:

N . . . G02/G03 X. . . . Z. . . . I. . . . K. . . .

Where G02 is clockwise circular interpolation, and G03 is counterclockwise circular interpolation; X is the X axis endpoint of the arc; Z is the Z axis endpoint of the arc; I is the X axis coordinate of the arc centerpoint; and K is the Z axis coordinate of the arc centerpoint. When the center is specified using a radius, the R address is used. R is programmed as an incremental value from the current tool position. The format is:

N . . . G02/G03 X. . . . Z. . . . R. . . .

Two programs are presented here for turning a spherical end on a 2.000-inch-diameter piece of 304 stainless steel (see Figure 14-8). Figure 14-9A is a program to turn the end using I and K; Figure 14-9B is identical except that R is used instead.

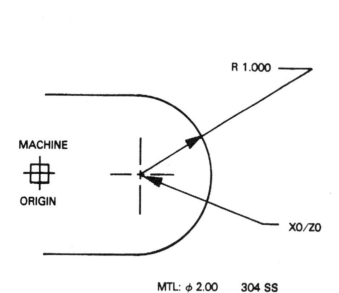

R 1.000

MACHINE

ORIGIN

X0/Z0

MTL: φ 2.00   304 SS

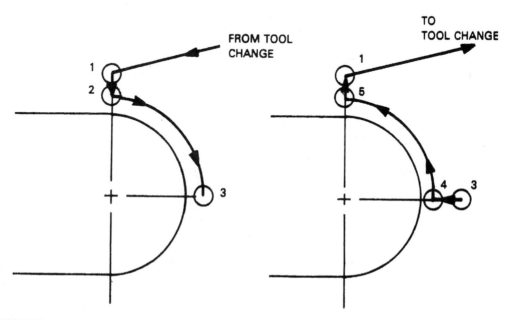

FROM TOOL
CHANGE

TO
TOOL CHANGE

FIGURE 14-8
Turning a spherical end

```
%
O1409
(* **********)
(* INCH PROGRAM - CIRCLE CENTER METHOD)
(*X0/Z0 = CENTERLINE OF PART RADIUS)
(* **********)
N010G00G99M08                    (SAFETY LINE)
N020T0101M42                     (TURRET POS, HIGH RANGE)
N030S150M03                      (SPINDLE ON)
N040X2.1Z0.                      (POSITION TO #1)
N050G01X2.084F.003               (FEED TO #2)
N060G02X0.Z1.042I0.K0.           (CW ARC TO #3)
N070G01Z1.032                    (FEED TO #4)
N080G03X2.062Z0.I0.K0.           (CCW ARC TO #1)
N090G00X2.084M09                 (RAPID TO #1, COOLNT OFF)
N100G28U0.W0.M05                 (RETURN TO HOME)
N110M30
%
```

```
%
O1409
(* **********)
(*INCH PROGRAM - RADIUS METHOD)
(*X0/Z0 = CENTERLINE OF PART RADIUS)
(* **********)
N010G00G99M08                    (SAFETY LINE)
N020T0101M42                     (TURRET POS, HIGH RANGE)
N030S150M03                      (SPINDLE ON)
N040X2.1Z0.M03                   (POSITION TO #1)
N050G01X2.084F.003               (FEED TO #2)
N060G02X0.Z1.042R1.042           (CW ARC TO #3)
N070G01Z1.032                    (FEED TO #4)
N080G03X2.062Z0R1.032            (CCW ARC TO #1)
N090G002.084M09                  (RAPID TO #1, COOLNT OFF)
N100G28U0.W0.M05                 (RETURN TO HOME)
N110M30
%
```

FIGURE 14-9
Program to turn part in Figure 14-8

Program Explanation (Refer to Figure 14-9)

N010

Safety line to cancel any active codes. Turns coolant on.

N020

T0101 – Selects tool #1, offset #1.

M42 – Selects high gear range.

N030

S150 – Sets the spindle speed to 150 RPM.

M03 – Turns the spindle on.

N040

X2.1 Z0. – Positions the tool to location #1, Figure 14-8.

N050

G01 – Selects feed rate movement.

X 2.084 – Feeds the tool from location #1 to location #2.

F.003 – Assigns the feedrate.

N060

G02 – Selects clockwise circular interpolation.

X0.Z1.042 – Arc endpoint coordinates, location #3.

10.K0. – Centerpoints of the arc, Figure 14-9, top.

R1.042 – Radius value, Figure 14-9, bottom. The 1.042 value is the incremental distance from the arc start point (location #2) to the arc center.

N070

G01 – Selects feedrate movement.

Z1.032 – Feeds the tool from location #3 to location #4.

N080

G03 – Selects counterclockwise circular interpolation.

X2.062 Z0.– Endpoint coordinates of the arc.

10.K0. – Centerpoints of the arc, Figure 14-9, top.

R 1.032 – Radius of the arc, Figure 14-9, bottom.

N090

G00 – Selects rapid traverse.

X2.084 – Rapids the cutter from location #5 to location # 1.

M09 – Turns the coolant off.

N100

G28U0.W0. – Returns the tool to the reference point.

M05 – Turns the spindle off.

N110

M30 – Signals end of program.

DRILLING

Drilling on NC lathes is accomplished in similar manner to turning and boring. The tool is sent to a desired start position, and the coordinates are given to move along the proper path. When drilling, the tool point is programmed since there is no tool radius involved. Canned cycles like those used for drilling on NC mills will be discussed in a later section.

To drill a ³/₈ diameter hole 1.500 inches deep in part Figure 14-1, a centerdrill and a ³/₈ drill can be added to the program in Figure 14-3. This has been done in Figure 14-10. The program explanation follows (see page 300).

```
%
O1410
(* *********)
(* X0 = CENTERLINE OF SPINDLE)
(* Z0 = PART SHOULDER )
(* *********)
N010G00G99M08
N020T0101M42
N030S1200M03
N040X2.6Z2.042
N050G01X0.F.007
N060Z2.032
N070X2.314F.003
N080Z.042F.007
N090X2.6
N100G00X2.32Z2.132
N110G01X2.084
N120Z.042F.003
N130X2.6
N140G00X2.084 Z2.132
N150G01X2.062
N160Z.032F.003
N170X2.55
N180G00U0.W0.M09
N190M01                          (OPSTOP)
(* *********)
(* C'DRILL)
(* *********)
N200M08                          (COOLNT ON)
N210T0202M42                     (TURRET POS & HIGH RANGE)
N220S1800M03                     (SPNDL ON, 1800 RPM)
N230G00X0.Z2.1                   (POSITION TO START)
N240G01Z-1.85F.003               (FEED TO DEPTH)
N250G00Z2.1                      (RAPID TO START POS.)
N260G28U0.W0.M09                 (RETURN TO REF, COOLNT OFF)
N270M01                          (OPSTOP)
(* *********)
(* DRILL)
(* *********)
N280M08                          (COOLNT ON)
N290T0303 M42                    (TURRET POS & HIGH RANGE)
N300S1600M03                     (SPNDL ON, 1600 RPM)
N310G00X0.Z2.1                   (RAPID TO START POS.)
N320G01Z1.625F.003               (FEED TO 1ST PECKING DEPTH)
N330G00Z2.5                      (RAPID OUT OF PART)
N340Z1.63                        (RAPID TO START OF PECK)
N350G01Z1.375                    (FEED TO 2ND PECKING DEPTH)
N360G00Z2.5                      (RAPID OUT OF PART)
N370Z1.38                        (RAPID TO START OF PECK)
N380G01Z1.                       (FEED TO 3RD PECKING DEPTH)
N390G00Z2.5                      (RAPID OUT OF PART)
N400Z1.005                       (RAPID TO START OF PECK)
N410G01Z.625                     (FEED TO 4TH PECKING DEPTH)
N420G00Z2.5                      (RAPID OUT OF PART)
N430Z.63                         (RAPID TO START OF PECK)
N440G01Z.387                     (FEED TO FINISH DEPTH)
N450G00Z.1                       (RAPID TO START POSITION)
N460G28U0.W0.M09                 (RETURN TO REF, COOLNT OFF)
N470M05                          (SPNDL OFF)
N480M30                          (END PRGM)
%
```

FIGURE 14-10

Program to machine part in Figure 14-1

Program Explanation

N010–N180 are identical to Figure 14-3.

N190

Optional stop code. This code aids the operator during setup. If the optional stop switch is turned on at the console, the program will stop at this line. The operator can then inspect the workpiece during setup. It is common practice to include an M01 at the end of each tool.

N200–N220

Selects the tool, offset, gear range. Turns the spindle and coolant on.

N230

G00 – Rapid traverse mode.

X0.Z2.1 – Rapids the centerdrill to the start position, .100 away from the workpiece face.

N240

G01 – Feedrate mode.

Z–1.85 – Depth of centerdrilling (.150 deep).

F.003 – Sets feedrate at .003 IPR.

N250

G00 – Rapid traverse mode.

Z2.1 – Returns tool to the start position.

N260

Returns tool to the reference point and cancels the tool offset.

N270

M01 – Optional stop code.

N280–N300

Selects tool, offset, gear range. Turns on spindle and coolant.

N310

Rapids tool tip to the start point.

N320

G01 – Feedrate mode.

Z1.625 – Depth of first drill peck.

F.003 – Sets the feedrate to .003 IPR.

N330

G00 – Rapid traverse mode.

Z2.5 – Sends the tool tip .500 away from the part face. The .500 distance gives the coolant sufficient area to enter the section of hole just drilled to lubricate the drill point on the next drill peck.

N340

Z1.63 – Sends the tool tip to the start of the next peck, .005 from the end point of the previous drill peck.

N350

G01 – Feedrate mode.

Z 1.375 – End point to the second drill peck.

N360

Rapids tool .500 out of part.

N370

Rapids tool tip to start of third peck.

N380–N440

The pecking cycle is repeated until final hole depth is achieved.

N450

Tool rapids out of part to original start position.

N460

Returns to reference line.

N470

Spindle off.

N480

END of program.

THREADING

When threading on CNC lathes, one of three threading cycles is used: single pass threading (G33), multiple pass threading (G92), or multiple pass threading (G76). When a G33 is issued, the tool travels the length of the thread and stops. The tool then has to be retracted from the thread, returned to the starting point, and the whole procedure repeated. When a G92 command is issued, the tool moves to a programmed X coordinate, feeds across the length of the thread to the programmed Z coordinate, and returns to the start point. This process is automatically repeated with the X-axis moving to a new programmed X coordinate until the final X coordinate has been executed. When a G76 is issued, the machine makes a threading pass, then automatically retracts the tool to the X axis reference position and returns it to the Z axis start position. It then automatically repeats the procedure until the final depth of the thread is achieved.

Three types of threads can be cut using a CNC lathe: constant lead, increasing lead, and decreasing lead. The lead of a thread is the distance that the thread advances in one revolution. Some CNC lathes are capable of cutting only constant lead threads, depending on the thread-cutting options selected when the machine is purchased. Threads of increasing and decreasing lead are specialized applications and will not be dealt with in this text.

When cutting threads, the relationship between spindle speed and tool feedrate is very important. When a G code is used for thread cutting, the feedrate override controls on the MCU console, which allow the operator to adjust the feedrate during machining, will not function. When beginning a threading pass, a certain distance (A in Figure 14-11) must be allowed ahead of the part face, to give the lathe carriage time to accelerate to the proper feedrate. Failure to allow this distance will result in improper leads on the first several threads.

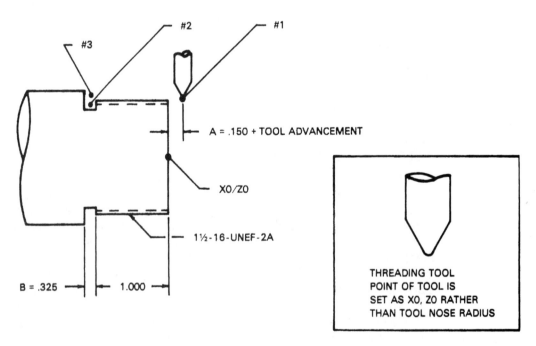

FIGURE 14-11
Part to be threaded

Starting distance A varies from machine to machine. Charts giving the distance for a particular thread on a particular machine will be found in the programming manual. If a chart is not available, the following formula can be used:

$$A = (RPM \times LEAD \times .006) + Z$$

Where Z is the amount of tool advancement in the Z axis. *Tool advancement* occurs, prior to the start of a threading cut, along two axes, as illustrated in Figure 14-12. Advancement along the Z axis is calculated by the formula:

$$Z = X \ (TAN \ 30)$$

Some programmers prefer to feed the tool in at a 29-degree angle instead of 30. In this case the formula would be:

$$Z = X \ (TAN \ 29)$$

The stopping distance is similar to the starting distance. This distance is shown in Figure 14-11 as dimension B. The minimum stopping distance can be calculated by the following formula if a chart is not available:

$$B = RPM \times LEAD \times .013$$

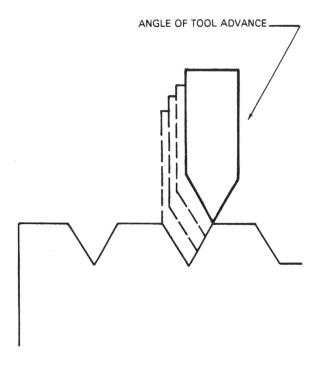

ANGLE OF TOOL ADVANCE

FIGURE 14-12
Tool advancement

Three threading programs have been written for the part shown in Figure 14-10. The program in Figure 14-13 cuts the thread using single pass threading. The program in Figures 14-14 and 14-15 cut the thread using multiple pass threading. The format for single pass threading is:

n . . . G33 . . . Z . . . . F . . . .

N . . . G33 . . . . F . . . .   (absolute positioning)

or

N . . . G33 U . . . . W . . . .   (incremental positioning)

On FANUC-style lathe controllers, G90 and G91 are not used to switch between absolute and incremental positioning. Instead, a secondary set of axes are used to specify incremental movement. The U axis specifies incremental motion along the lathe's X axis. The W axis specifies incremental motion along the lathe's Z axis.

```
%
O1413
(* **********)
(* X0 = SPINDLE CENTERLINE)
(* Z0 = PART FACE)
(* **********)
N0101G00G99M08
N0202T0101M42
N030S400M03
N040X1.47Z.015                     (POSITION TO #1)
N050G91G33W-1.15F.0625             (1ST THREAD PASS)
N060G00U.015                       (RETRACT XAXIS)
N070W1.168                         (RETURN ZAXIS TO START)
N080U-.032W-.018                    (ADVANCE TOOL)
N090G33W-1.168F.0625               (2ND THD PASS)
N100G00U.032                       (RETRACT XAXIS)
N110W1.186                         (RETURN ZAXIS TO START)
N120U.032W-.018                    (ADVANCE TOOL)
N130G33W-1.186F.0625               (3RD THREAD PASS)
N140G00U.032M09                    (RETRACT XAXIS)
N150G28U0.W0.M05                   (RETURN AXES TO HOME)
N160M30
%
```

FIGURE 14-13
Thread program using G33 thread cycle

```
%
O1414
(* **********)
(* X0 = SPINDLE CENTERLINE)
(* Z0 = PART FACE)
(* **********)
N010G0G99M08
N020T0101M42
N030S700M03
N040G00X1.6Z.15          (THD. START POINT)
N050G92X1.58Z-1.15       (1ST PASS)
N060X1.57                (2ND PASS)
N070X1.55                (3RD PASS)
N080X1.53                (4TH PASS)
N090X1.51                (5TH PASS)
N100X1.49                (6TH PASS)
N110X1.47                (7TH PASS)
N120X1.46                (8TH PASS)
N130X1.455               (9TH PASS)
N140X1.45                (10TH PASS)
N150X1.445               (11TH PASS)
N160X1.443               (12TH PASS)
N170X1.44                (13TH PASS)
N180X1.438               (4TH PASS)
N190X1.437               (15TH PASS)
N200X1.436               (16TH PASS)
N210G28U0.W0.M09
N220M05
N230M30
%
```

FIGURE 14-14
Thread program using G92 thread cycle

```
%
O1415
(* **********)
(* X0 = CENTERLINE OF SPINDLE)
(* Z0 = PART FACE)
(* **********)
N010G00G99M08
N020T0101M42
N030S400M03
N040X1.6Z.15                              (THREAD START POINT)
N050G76X1.436Z1.I0.K.032F.0625D.015A60    (THREADING CYCLE)
N060G00G28U0.W0.M09
N070M05
N080M30
%
```

FIGURE 14-15
Thread program using G76 thread cycle

Where G33 is the thread-cutting G code, Z is the length of the threading cut, and F is the lead of the thread. (Some lathe controllers use K to specify the lead of the thread.)

The format for G92 multipass threading is:

N . . . G92 X. . . . Z. . . . F. . . .
N . . . X . . .
N . . . X . . .

       .

       .

       .

N . . . X . . .

Where: G92 = multipass threading code
           X = X coordinate of the first threading pass
           Z = Z coordinate of the threading end point
           F = the feedrate (lead) of the thread

           X = depth of second pass
           X = depth of third pass

    etc. until

           X = depth of final pass

Usually the lead can be given to only four decimal places, so that some round-off error will occur. This is usually so slight that it will affect only threads several feet long. Some machines have the capacity to accept thread leads to five or six decimal places.

The format for G76 multiple pass threading is:

N . . . G76 X. . . . Z. . . . I. . . . K. . . . D. . . . F. . . . A. .

Where: G76 = multipass threading G code
           X = minor diameter of the thread
           Z = length of thread
           I = difference in thread radius from one end of the thread to the other. This value is used for cutting tapered threads. For straight threads, a value of zero is entered.
           K = height of the thread (a radius value, given from the crest of the thread to the root)
           D = depth of cut for the first pass
           F = lead of the thread
           A = angle of the tool tip. (For Unified, American National, and IFI metric threads, the angle is 60 degrees.)

The main differences between G92 and G76 are first, G76 requires only one line of program code, and second, G92 plunges the tool straight into the workpiece rather than feeding in at an angle. The infeed direction with G76 is 30 degrees. The infeed direction with G33 is controlled by the programmer.

Program Explanation (Refer to Figure 14-13)

N010

Safety line, returns tool to reference.

N020

M06 T0101 – Selects tool #1, offset #1.

N030

S400 – Sets the spindle speed to 400 RPM.
M03 – Turns on the spindle.

N040

X1.47 Z.15 – Coordinates of location #1, Figure 14-11. The X coordinate is diameter programmed and positions the tool to the depth of the first pass. The Z coordinate is the starting distance. Subsequent passes will add to the starting distance the amount of Z-axis tool advancement.

N050

G91 – Selects incremental positioning.
G33 – Initiates single pass threading.
W1.15 – Feeds the tool from location #1 to location #2, Figure 14-11.
F.0625 – Lead of the thread.

N060

G00 – Selects rapid traverse.
U.015 – Incrernental coordinate to rapid the tool from location #2 to location #3.

N070

W1.168 – Incremental distance to rapid the tool back to the starting point. This coordinate also compensates for the additional starting distance required by the tool advancement for the next pass.

N080

U-.032 – Incremental coordinate to advance the tool for the next cut. Two .015-inch roughing cuts are being made. This coordinate advances the X axis the .015 inch the tool was retracted at the end of the first pass, plus the .015 inch desired for the second.

W-.018 – Calculated Z-axis tool advancement to cause the tool to advance on a 30-degree angle.

N090

G33 – Initiates the threading cycle.
W-1.168 – Feeds the tool from the start point (location #1) to the end of the thread point (location #2).
F.0625 – Lead of the thread.

N100

G00 – Selects rapid traverse.
U.032 – Retracts the X axis from the thread.

N110
W1.168 – Returns the tool to the starting point of the thread.

N120
U.032 W-.018 – Advances the tool to final thread depth.

N130
G33 – Initiates thread cutting.
W-1.168 – Feeds the tool from #1 to #2.
F.0625 – Lead of the thread.

N140
G00 – Selects rapid traverse.
U.032 – Retracts the tool from the thread.
M09 – Turns off the coolant.

N150
G90 – Selects absolute positioning.
G28U0.W0. – Returns the tool to the reference point.
X6 Z6 – Intermediate point coordinates.
M05 – Turns off the spindle.

N160
M30 – Signals end of program.

Program Explanation (Refer to Figure 14-14)

N010
Safety line, returns to reference.

N020
T0101 – Selects tool and offset.
M42 – Selects high gear range.

N030
S700 M03 – Turns the spindle on at 700 RPM.

N040
X1.6 Z.15 – Start position of the thread.

N050
G92 – Initiates threading cycle.
X1.58 – X coordinate of first threading pass.
Z-1.15 – Z coordinate of the ending point.
F.0625 – The thread lead.

N060–N200
X coordinates of the succeeding thread passes. N200 is the last pass. Note the passes gradually remove less and less stock per pass to eliminate tearing of the thread.

N210–N220

Returns the tool to reference. Turns off coolant and spindle.

N230

END of program.

Program Explanation (Refer to Figure 14-15)

N010

Safety line.

N020

T0101 – Selects tool #1, offset #1.

N030

S400 – Sets the spindle speed.

M03 – Turns the spindle on.

N040

Z1.5 – Positions the Z axis at the start of the thread.

G76 – Initiates multipass threading.

X1.436 – Minor diameter of the thread.

Z1 – Length of the thread.

10 – Difference in radius of the thread from the starting point to the finish point.

K.032 – Height of the thread measured from the crest to the root.

D.015 – Specifies a .015-inch first pass.

F.0625 – Lead of the thread.

A60 – Specifies a 60-degree thread.

N060

G00 – Selects rapid traverse.

G28U0.W0. – Initiates a return to reference.

M09 – Turns off the coolant.

N070

M05 – Turns off the spindle.

N080

M30 – Signals the end of program.

Note how the amount of programming is reduced when using the multipass cycle.

Do loops and subroutines may also be used in lathe programming; they are programmed in just as when milling. Tool nose radius compensation may also be used. TNR comp has not been discussed here in order to concentrate on the basics of tool nose centerline programming. It is used in similar fashion to cutter diameter compensation in CNC milling programs. Once tool nose centerline programming is understood, there should be no problem in using TNR comp. The same ramp on/ramp off precautions apply in turning as in milling.

A COMPLETE LATHE EXAMPLE

Up to this point, small lathe programming routines have been presented. These routines illustrate various lathe operations which usually are parts of a single lathe program. Figure 14-16 is a part for which a program has been written. The program is contained in Figure 14-17. A brief program explanation follows. There are several codes used in this program that should be noted.

G98—used to select inch per revolution feedrates.
G97—used to select direct RPM programming.
M24—used when threading to cause the tool to pull straight out of the part. The default condition for a thread cycle is for the tool to pull out at a 60 to 45 degree angle.

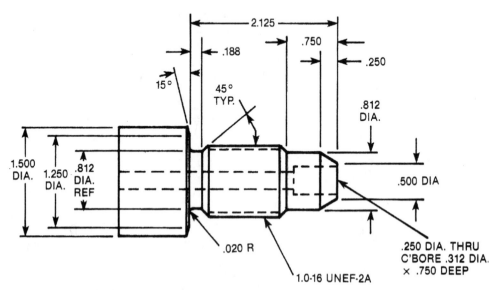

FIGURE 14-16
Part drawing

```
%
O1417
(* *********)
(* LATHE PROGRAMMING EXAMPLE)
(* X0 = CENTERLINE OF PART)
(* Z0 = FACE OF PART)
(* *********)
(*.031R X 80 DEG. TURNING TOOL)
(* *********)
N1G97
N2G99
N3M08
N4G00T0101
N5S2133M03
(ROUGH FACE PART - LEAVE .005 STK.)
N6X1.Z.031
N7G01X0.F.007
N8G00Z.1
( ROUGH TURN 1.0 DIA. IN 2 PASSES - LEAVE .005 STK./SIDE)
N9X1.172
N10G01Z-2.089F.0070
N11X1.672
N12G00Z.1
N13X1.072
N14G01Z-2.089F.007
N15X1.2594
N16X1.672Z-2.1443
N17G00Z.136
N18G28U0.W0.
N19M01
(* *********)
(* .007R X 35 DEG. TURNING TOOL)
(* *********)
N20G99
N21M08
N22G00T0202
N23S2133M03
(ROUGH THREAD RELIEF AREA)
N24X1.074Z-1.823
N25G01X.836Z-1.9420F.003
N26Z-2.105
N27G02X.852Z-2.113I.852K-2.105
N28G01X1.084
( FINISH O.D.)
N29G00Z.007
N30G01X0.F.003
N31X.489
N32G03X.5178Z-.001I.4889K-.01
N33G01X.8208Z-.2439
N34G03X.826Z-.2529I.792K-.2529
N35G01Z-.7471
N36X1.004Z-.8361
N37G03X1.014Z-.8481I.98K-.8481
N38G01Z-1.8389
```

FIGURE 14-17 *(Continues to page 314)*

```
N39G03X1.004Z-1.8509I.98K-1.8389
N40G01X.826Z-1.9399
N41Z-2.105
N42G02X.852Z-2.118I.852K-2.105
N43G01X1.2518
N44X1.614Z-2.1665
N45G00Z.1070
N46G28U0.W0.
N47M01
(* **********)
(* THREADING TOOL)
(* THREAD O.D. 1-16-2A)
(* **********)
N48G99
N49M08
N50G00T0303
N51S900M03
N52X-.5Z.6M74
N53G92X.99Z-2.1F.0625
N54X.98
N55X.9718
N56X.9654
N57X.96
N58X.9552
N59X.951
N60X.947
N61X.9434
N62X.94
N63X.9368
N64X.9336
N65X.9308
N66X.9278
N67X.9252
N68X.9234
N69X.9234Z-2.1
N70G28U0.W0.
N71M01
(* **********)
(* NO. 4 C'DRILL)
(* C'DRILL TO .260 DIA.)
(* **********)
N72G99
N73M08
N74G00T0404
N75S3000M03
N76X0.Z.1
N77G01Z-.278F.003
N78G00Z.1
N79G28U0.W0.
N80M01
```

```
(* **********)
(* 1/4 DRILL)
(* DRILL .250 DIA. THRU)
(* **********)
N81G99
N82M08
N83G00T0505
N84S2000M03
N85X0.Z.1
N86G01 Z-.3F.003
N87G00Z.5
N88Z-.295
N89G01Z-.6F.003
N90 G00 Z.5
N91-.595
N92G01Z-.9F.003
N93G00Z.5
N94Z-.895
N95G01Z-1.2F.003
N96G00Z.5
N97Z-1.195
N98G01Z-1.5F.003
N99G00Z.5
N100Z-1.495
N101G01Z-1.8F.003
N102G00Z.5
N103Z-1.795
N104G01 Z-2.1F.003
N105G00 Z.5
N106Z-2.095
N107G01Z-2.4F.003
N108G00Z.5
N109Z-2.395
N110G01 Z-2.7F.003
N111G00 Z.5
N112Z-2.695
N113G01Z-3.F.003
N114G00Z.5
N115Z-2.995
N116G01Z-3.25F.003
N117G00Z.1
N118G28U0.W0.
N119M01
(* **********)
(* .005R BORING BAR)
(* **********)
N120G99
N121M08
N122G00T0606
N123S3500M03
(ROUGH C'BORE - LEAVE .005 STK/SIDE)
N124X.292Z.035
N125G01Z-.74F.002
N126X.152
```

```
N127G00Z.04
(FINISH C'BORE - DEBURR EDGE WITH .01R)
N128X.332
N129G01Z.0105F.002
N130G02X.292Z-.0095I.332K-.0095
N131G01Z-.74
N132X.132
N133G00Z.11
N134G28U0.W0.M09
N135M05
N136M30
%
```

FIGURE 14-17
Program to machine the part in Figure 14-16

Program Explanation

First Tool:

NI–N5
Selects first tool. Tums on spindle and coolant.

N6–N8
Part is rough faced with .005 stock left for finishing.

N9–N11
First roughing pass on o.d.

N12–N17
Second roughing pass on o.d. The 15 degree angle is also rough turned at this time.

N18
Tool is returned to reference point. Tool offset cancelled.

Second Tool:

N20–N23
Selects second tool. Turns on spindle and coolant.

N24–N28
Thread relief area is rough turned. .005 stock is left for finishing.

N29–N31
Face of part is finished.

N32
Deburring radius is turned at the intersection of the first angle and the face of the part.

N33

First angle is finish turned.

N34

Deburring radius is turned at the intersection of the first angle and the .812 diameter.

N35

The .812 diameter is finish turned.

N36

The front thread chamfer is finish turned.

N37

A radius is turned at the intersection of the thread chamfer and major diameter.

N38

The major diameter of the thread is turned.

N39

A radius is turned at the intersection of the back thread chamfer and major diameter.

N40

The back thread chamfer is turned.

N41

The .812 diameter thread relief is turned.

N42

The .020 radius is turned.

N43

The 2.125 dimension is faced.

N44

The 15 degree angle is finish turned.

N45–N47

The tool is returned to reference. The offset is cancelled.

Third Tool:

N48–N51

Tool and offset selected, spindle and coolant turned on.

N52

The tool is sent to the start position for threading.
The M74 turns off the thread chamfering at the end of a thread pass.

N53

G92 multi-pass thread cycle initiated.

N54–N68
Succeeding X values for the G92 cycle. Each X value is used on a separate thread pass.

N69
Last threading pass which is a repeat pass. The Z coordinate is optional.

N70–N71
Returns the tool to reference. Offset is cancelled.

Fourth Tool:

N72–N75
Tool, offset, spindle speed selected.

N76–N78
Drill sent to start point, fed to depth, and rapids back to start position.

N79–N80
Returns to reference.

Fifth Tool:

N81–N84
Tool, offset, spindle speed selected.

N85–N116
Peck drilling of 1/4 inch through hole. Each peck is .300 deep. At end of peck the tool is sent at rapid z.500 to clear out chips and allow coolant into the hole. The tool sequence repeats until final depth is achieved in N116.

N117
Tool is returned to the starting position.

N118–N119
Returns to reference.

Sixth Tool:

N120–N123
Tool, offset, spindle speed selected.

N124–N127
The c'bore is rough bored. .005 stock is left for finishing.

N128–N130
A deburring radius is turned at the intersection of the c'bore and the part face.

N131–N133
The c'bore is finish bored and tool retracted from part.

N134
Return to reference line, coolant off.

N135
Spindle off.

N136
END of program.

CANNED CYCLES

Most modern CNC lathe controllers contain a number of built-in canned cycles. The threading cycles G33, G92, and G76 are standard from controller to controller. Other canned cycles are options offered by the controller manufacturer. These cycles are often unique to a given controller manufacturer (sometimes unique to a given model of controller) and therefore not transportable from controller to controller. With the current CNC lathe investment strategies by small and midsized companies, canned cycles will become as standardized as mill cycles at some future point. It is not possible to cover the number of cycle variations in a text of this size. The student should be aware, however, that these cycles exist. Documentation on the use of these cycles will be contained in the programming and operational manuals for a given machine.

How much a company relies on canned cycles for lathe programming is dependent on their use or non-use of computer-aided programming. Where computer-aided or graphics programming is utilized, there is little need for canned cycles aside from the standard lathe threading cycles. Where MDI programming is used, canned cycles can save many hours of programming time. The cycles used in these situations usually include: rough turning and boring cycle, rough facing cycle, finish turning and boring cycle, finish facing cycle, peck drilling cycle, step drilling cycle, chamfering cycle, and growing cycle.

One caution should be noted by the programmer: Canned cycles valid for one controller can cause a crash situation if run on an incompatible controller if the controller does not stop and put out an alarm message when the canned cycle is encountered.

Appendix 7 contains a sample program utilizing canned cycles, along with an explanation of how the cycles are used.

SUMMARY

The important concepts presented in this chapter are:

- In diameter programming, the X-axis coordinates are one-half the actual tool movement.
- In radius programming, the X-axis coordinates and the tool movement are the same.

- G01, linear interpolation, is used for feedrate moves.
- Coordinates for taper turning must be calculated using trigonometry (or other math methods), just as when milling angles.
- G02 and G03 are used for circular interpolation.
- I and K are the addresses used to program the center points of an arc.
- The R address is used in place of I and K to program an arc using the arc radius instead of the arc centerpoints.
- Single pass threading cycles produce one threading cut. The cycle must be reinitiated for each threading pass.
- Multipass threading can produce an entire finished thread without additional programming.
- When threading, the Z axis tool advance must be calculated from the X-axis depth of cut by the formula Z = X TAN(30).
- Minimum starting and stopping distances must be calculated for use in a threading program.

VOCABULARY INTRODUCED IN THIS CHAPTER

Constant lead thread
Decreasing lead thread
Diameter programming
Increasing lead thread
Intermediate point
Radius programming
Reference point

REVIEW QUESTIONS

1. What G codes are used for feedrate moves?
2. What is the difference between diameter and radius programming?
3. Write a program to turn and face the part in Figure 14-18.
4. Write a program to turn the taper on the part in Figure 14-19.
5. What codes are used to institute circular interpolation?
6. What addresses are used to define the X and Z axis center point of an arc?
7. What address is used to define the arc using the arc radius?
8. Write a program to machine the part in Figure 14-20.
9. What is the code for single pass threading? What is the format?
10. What is the code for multipass threading? What is the format?
11. Write a program to thread the part in Figure 14-21:
    a. Using single pass threading.
    b. Using multipass threading G92.
    c. Using multipass threading G76.

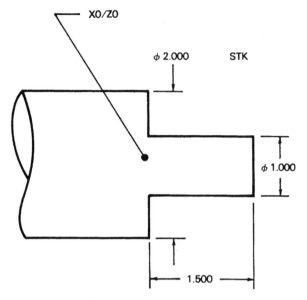

MATERIAL: CARPENTER STENOR TOOL STEEL

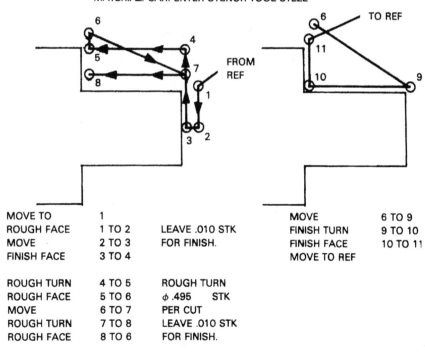

MOVE TO	1		MOVE	6 TO 9
ROUGH FACE	1 TO 2	LEAVE .010 STK	FINISH TURN	9 TO 10
MOVE	2 TO 3	FOR FINISH.	FINISH FACE	10 TO 11
FINISH FACE	3 TO 4		MOVE TO REF	

ROUGH TURN	4 TO 5	ROUGH TURN
ROUGH FACE	5 TO 6	ϕ .495 STK
MOVE	6 TO 7	PER CUT
ROUGH TURN	7 TO 8	LEAVE .010 STK
ROUGH FACE	8 TO 6	FOR FINISH.

FIGURE 14-18
Part drawing for review question #3

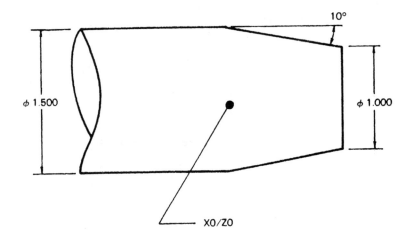

MATERIAL: ϕ 1.500   4140 STEEL

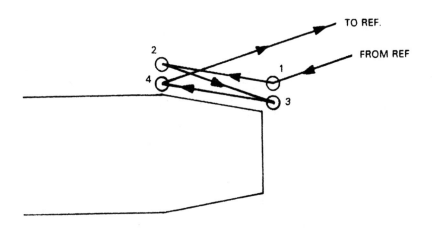

MOVE TO 1
ROUGH TURN   1 TO 2
MOVE   2 TO 3
FINISH TURN   3 TO 4
MOVE   TO REF

FIGURE 14-19
Part drawing for review question #4

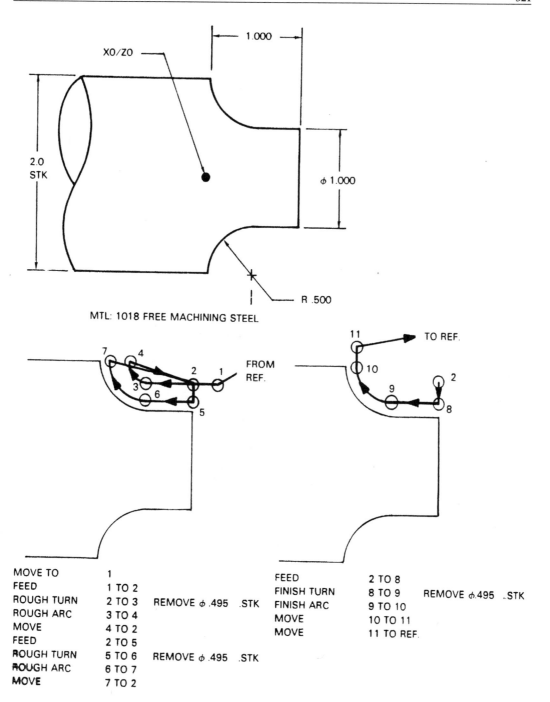

MTL: 1018 FREE MACHINING STEEL

MOVE TO	1		FEED	2 TO 8	
FEED	1 TO 2		FINISH TURN	8 TO 9	REMOVE φ.495 .STK
ROUGH TURN	2 TO 3	REMOVE φ.495 .STK	FINISH ARC	9 TO 10	
ROUGH ARC	3 TO 4		MOVE	10 TO 11	
MOVE	4 TO 2		MOVE	11 TO REF.	
FEED	2 TO 5				
ROUGH TURN	5 TO 6	REMOVE φ.495 .STK			
ROUGH ARC	6 TO 7				
MOVE	7 TO 2				

FIGURE 14-20
Part drawing for review question #8

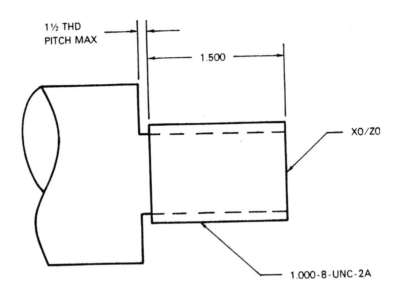

1½ THD
PITCH MAX

1.500

X0/Z0

1.000-8-UNC-2A

MATERIAL: 4130 STEEL

FIGURE 14-21
Part drawing for review question #11

CHAPTER 15

Use of Computers in Numerical Control Programming

OBJECTIVES Upon completion of this chapter, you will be able to:

- Describe the three basic ways computers are used in numerical control programming.
- Describe offline programming.
- Explain the advantages of computer-aided programming.
- Describe the three types of statements used in a computer-aided program.
- Understand basic part geometry and tool motion statements in APT and COMPACT II.
- Describe two types of computer graphics programming systems and how they differ.
- Explain how graphics programming simplifies writing NC programs.

Computers are becoming commonplace on shop floors. This is particularly true in numerical control programming. Only a few years ago, computers for NC programming were limited to large companies, but with the advent of inexpensive computer memory, even the smallest mold shops can now afford them. Computers can be used to help write NC programs in three basic ways: in offline programming, computer-aided programming, and computer graphics programming. This chapter will introduce all three of these uses. It is not the purpose of this text to teach computer-aided or computer graphics programming but, rather, to make the student aware of the far-reaching effects that computers are having on manufacturing.

OFFLINE PROGRAMMING TERMINALS

Technically speaking, all uses of computers to assist programming are offline programming methods. *Offline programming is* programming that is performed away from the machine, not at the CNC computer keyboard. An offline programming terminal usually refers to a computer that is used as a text editor for writing programs. This type of programming station does not "aid" the programmer except by allowing the program to be entered into the computer exactly as if it were being entered via the MDI console. The advantage is that one program may be written while another program is being run on the machine. The program being written is simply saved on a computer disk, magnetic tape, punched tape, or a combination of these three.

COMPUTER-ASSISTED PROGRAMMING

As the name implies, computer-assisted programming involves the use of computers to help the programmer write an NC program. Computer-aided programming can be divided into two main categories: Language based programming, referred to simply as *Computer-aided programming,* and Computer Aided Manufacturing, called *CAM* programming. The difference lies primarily in the way the programming interacts with the programming system.

Computer-aided programming had its beginnings in the 1950s when point-to-point tape machinery made manual programming an enormously laborious task. A computer-aided programming system uses an English-like computer language to define a part's geometric configuration and describe the desired machining operations. The language acts as a translator. The programmer "talks" to the computer via the keyboard in the programming language. These language statements are stored in a file, called the source code file. The source code file is then processed into the required NC format by the computer. Originally these systems required large mainframe computers to handle the complicated calculations and processing involved. Today, however, language-based systems also run on desktop personal computers (PCs). The use of programming languages is diminishing due to the rising popularity of CAM programming systems. However, a number of large companies, particularly in the aerospace field, still use them.

CAM programming uses computer graphics, rather than a text based language, to interact with the programmer. The programming also can be graphically simulated by the computer. By "playing back" a cutting tool sequence, programming errors that otherwise might not be seen until the part is actually run on the machine, can be corrected during the programming process. Due to the technological advances in microelectronics, CAM systems, which run on PCs are the most common programming systems in use today.

Figure 15-1 illustrates the way a computer-aided programming system processes an NC program. Figure 15-2 illustrates the processing as performed by a CAM system. Each system takes the source code file(s) and processes them through a first-stage processor. The result of this initial processing is a special type of file called a centerline data file *(CL File)*. This CL file contains the centerline data for the cutter motion, and the auxiliary statements. Auxiliary statements contain such information as spindle speed, when to turn on the spindle, coolant commands, tool numbers, etc. The CL file is then run through another processing sequence, called *post-processing.* The postprocessor program takes the CL file data and converts it to the specific format required by a given machine. Since no two NC machines are alike, each NC machine requires its own postprocessor.

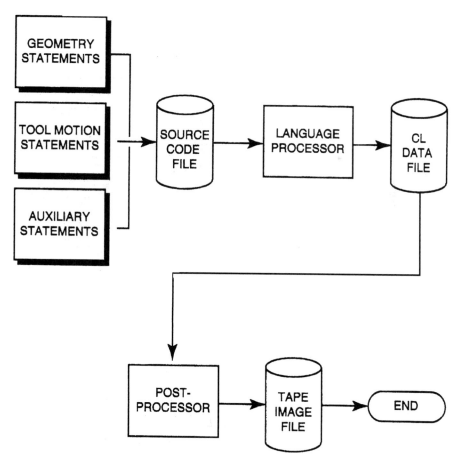

FIGURE 15-1
Computer-aided programming flow

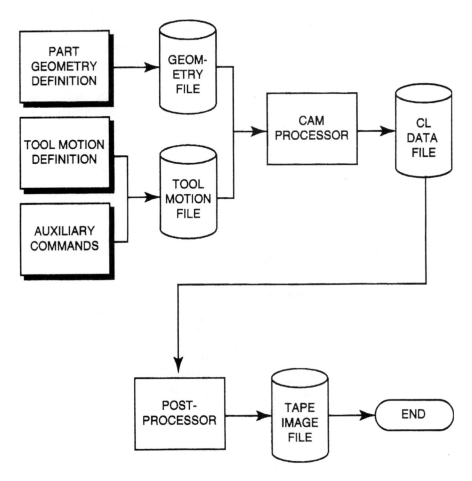

FIGURE 15-2
CAM programming flow

COMPUTER-AIDED PROGRAMMING LANGUAGES

The language a particular company uses depends on the parts it produces, the machines it uses, the cost of the programming system, and the computers that are available. Following are some of the more common programming languages.

APT (Automatic Programmed Tools).   APT is the oldest and the largest of the computer-aided programming languages. It can be used on only large computers and will perform the mathematical calculations required for complex curved surfaces using four and five axes.

AD-APT (Adaption of APT). This is a version of APT that can run on smaller computers. It uses about half the commands of APT and can be used for two-axis contouring with a third axis of linear motion.

AUTOMAP (Automatic Machining Program). AUTOMAP is another adaption of APT. It will run on medium size computers, has a limited number of commands, and is used for programming straight lines and circles.

COMPACT II. Compact II can accomplish the same tasks as APT but is limited to three-axis work. The main difference between them is that Compact II uses a somewhat more conversational set of commands, without some of the strict rules of syntax that must be observed in APT. The computer will also aid in debugging the finished program by interactive conversation with the programmer.

UNIAPT. UNIAPT is very similar to APT but is designed to run on dedicated minicomputers—that is, small computers that are used for only one task. UNIAPT will handle four- and five-axis programming.

NUFORM. NUFORM differs from most other computer-aided languages in that the programmer inserts codes or dimensions in their appropriate location within the NUFORM format. NUFORM uses numeric rather than alphabetic codes. In this chapter, APT and COMPACT II will be used as examples. An APT or COMPACT II program has three parts: part geometry definitions, auxiliary function statements (tool changes, speeds, feedrates, etc.), and tool motion statements.

Part Geometry Definitions in APT

Parts are defined to the computer by describing features such as points, lines, planes, circles, and their relationships to each other. Following are some of the simpler APT geometry commands.

To Define a Point. Points are the basis for defining other geometric features. Points may be defined by Cartesian coordinates or by relationship to other geometric features such as lines or circles. The following examples are all point definitions.

Point defined by Cartesian coordinates, Figure 15-3(a):
SYMBOL FOR POINT = POINT/X,Y,Z
Example: P1 = POINT/6,6,6
Meaning: P1 = a point X6.0000, Y6.0000, Z6.0000 from 0/0.

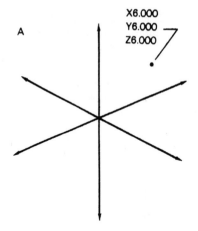

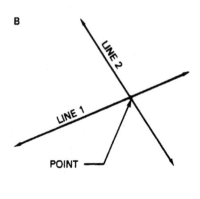

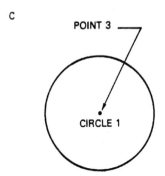

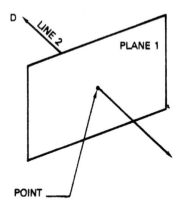

Defining a point in APT

FIGURE 15-3
Defining a point in APT

Point defined by the intersection of two lines, Figure 15-3(b):
SYMBOL FOR A POINT = POINT/INTOF, SYMBOL FOR A LINE, SYMBOL
    FOR A LINE
Example: P2 = POINT/INTOF,LN1,LN2
Meaning: P2 = a point located at the intersection of line LN1 and line LN2.

Point defined by the center of a circle, Figure 15-3(c):
SYMBOL FOR A POINT = POINT/CENTER, SYMBOL FOR A CIRCLE
Example: P3=POINT/CENTER,CIR1
Meaning: P3 = a point located at the center of circle CIR1.

Point defined by the intersection of a line and a plane, Figure 15-3(d):
SYMBOL FOR A POINT = POINT/INTOF, SYMBOL OF A PLANE, SYMBOL
    OF A LINE
Example: P4 = POINT/INTOF,PN2,LN3
Meaning: P4 = a point located at the intersection of line LN3 with plane PN2.

To Define a Line. Lines may be defined by using coordinates, other lines, points, and circles. Lines extend to infinity and are calculated by the computer mathematically from the information given it. It takes two points to define a line.

Lines defined by two points, Figure 15-4(a):
SYMBOL FOR A LINE = LINE/SYMBOL FOR A POINT, SYMBOL FOR A
    POINT
Example: LN1=LINE/PT1,PT2
Meaning: LN1 = a line passing through point PT1 and PT2.

Lines defined by a point and parallel line, Figure 15-4(b):
SYMBOL FOR A LINE = LINE/SYMBOL FOR A POINT, PARLEL, SYMBOL
    FOR A LINE
Example: LN2=LINE/PT2,PARLEL,LN1
Meaning: LN2 = a line passing through point PT2, parallel to line LN1.

Lines defined by a point and a perpendicular line Figure 15-4(c):
SYMBOL FOR A LINE = LINE/SYMBOL FOR A POINT, PERPTO, SYMBOL
    FOR A LINE
Example: LN3=LINE/PT3,PERPTO,LN2
Meaning: LN3 = a line passing through point PT3, perpendicular to line LN2.

Lines defined by a point and tangency to a circle, Figure 15-4(d):
SYMBOL FOR A LINE = LINE/SYMBOL FOR A POINT, LEFT/RIGHT, TANTO,
    SYMBOL FOR A CIRCLE
Example: LN4=LINE/PT2,LEFT,TANTO,CIR1
Meaning: LN4 is the line passing through point 2 and tangent to a circle with
    radius 1 on the left side.

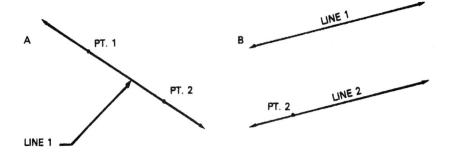

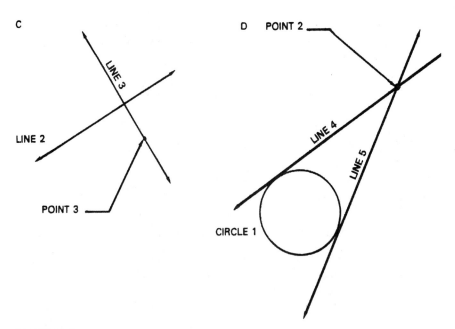

FIGURE 15-4
Defining lines in APT

Example: LN5 = LINE/PT3,RIGHT,TANTO,CIR1

Meaning: LN5 is the line passing through point 3 and tangent to a circle with radius 1 on the right side.

To Define a Circle. A circle is defined by an arc section. The section of the circle that is a part feature is programmed later as part of the tool motion statements.

Circle defined by the centerpoint and radius, Figure 15-5(a):

SYMBOL FOR A CIRCLE = CIRCLE/CENTER, SYMBOL FOR A POINT, RADIUS, RADIUS VALUE

Example: C1 =CIRCLE/CENTER,PT1,RADIUS,1.0

Meaning: C1 is a circle with center at point PT1 and a radius of 1.0 inch.

Circle defined by the centerpoint and a point on the circumference, Figure 15-5(b):

SYMBOL FOR A CIRCLE = CIRCLE/CENTER, SYMBOL FOR THE POINT AT THE CENTER, SYMBOL FOR THE POINT ON THE CIRCUMFERENCE

Exampe: C2=CIRCLE/PT2,PT4

Meaning: C2 is the circle with PT2 as its center and PT4 on its circumference.

Circle defined by the centerpoint and a tangent line, Figure 15-5(c):

SYMBOL FOR A CIRCLE = CIRCLE/SYMBOL FOR A POINT, TANTO, SYM-BOL FOR A LINE

Example: C3=CIRCLE/CENTER,PT5,LN1

Meaning: C3 is the circle with point 5 as its centerpoint, tangent to line 1.

To Define a Plane. Part surfaces are often defined by planes. Following are some simple definitions of planes.

Defining a plane by points, Figure 15-6(a):

SYMBOL FOR A PLANE = PLANE/SYMBOL FOR A POINT, SYMBOL FOR A POINT, SYMBOL FOR A POINT

Example: TOP = PLANE/PT1,PT2,PT3

Meaning: TOP is the plane passing through points 1, 2, and 3.

Defining a plane by a point and a parallel plane, Figure 15-6(b):

SYMBOL FOR A PLANE = PLANE/SYMBOL FOR A POINT, PARLEL, SYM-BOL FOR A PLANE

Example: BOTTOM=PLANE/PT4,PARLEL,TOP

Meaning: BOTTOM is the plane passing through point 4 and parallel with plane TOP.

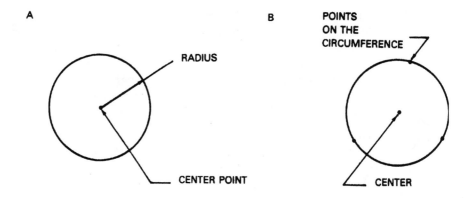

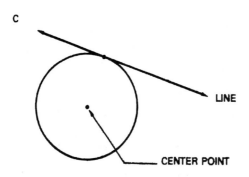

FIGURE 15-5
Defining circles in APT

Defining a plane through two points, perpendicular to another plane, Figure 15-6(c):

SYMBOL FOR A PLANE = PLANE/PERPTO, SYMBOL FOR A PLANE, SYMBOL FOR A POINT, SYMBOL FOR A POINT

Example: PLN3=PLANE/PERPTO,PLN2,PT3,PT5

Meaning: PLN3 is the plane perpendicular to plane PLN2, and passing through points PT3 and PT5.

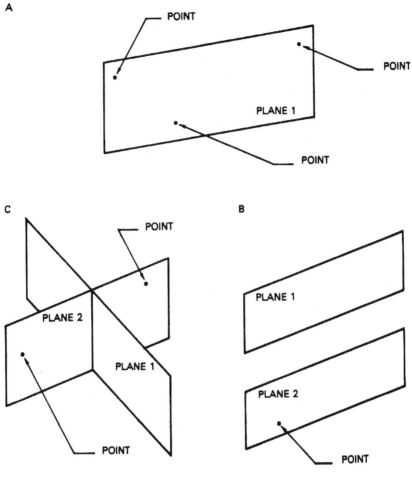

FIGURE 15-6
Defining planes in APT

Geometric definitions can also contain modifiers, which help in clarifying a definition. These modifiers are:

XLARGE   XSMALL
YLARGE   YSMALL
ZLARGE   ZSMALL

LARGE means that the side with the greater value is specified. SMALL specifies the side of lesser value. The value is determined by the coordinate position of the feature in question. The statement LN2 = LINE/
PARLEL,LN1,YLARGE,1.000 means that LN2 is line 2, parallel to line LN1 on
the side where the Y coordinates are the largest and 1.000 inch away from
LN1. YSMALL would mean that the line was 1.000 inch away on the side
where the Y coordinates were smaller.

Other types of features can be used to define the geometric shape of a
part, such as cylinders, vectors, surfaces, and angles. An APT dictionary, or a
textbook on computer-aided programming, will contain a comprehensive list
of APT geometric definitions.

Auxiliary Statements in APT

Before defining a cutter path using tool motion statements, the various
tools to be used must be defined to the computer. The tool statements are
auxiliary statements. The spindle speeds and feedrates used with the various
tools are also auxiliary statements. Following are examples of tool, machine,
spindle speed, and feedrate auxiliary statements.

Feedrates.

To define a feedrate in inches per minute:
FEDRAT/[feedrate value],IPM
Example: FEDRAT/20,1PM
Meaning: Feedrate is set to 20 inches per minute

To define a feedrate in inches per revolution:
FEDRAT/[feedrate value],IPR
Example: FEDRAT/.007,1PR
Meaning: Feedrate is set to .007 inch per revolution.

Tool Changes.

To change a tool when premeasured tools are not used:

LOADTL/[tool no.],ADJUST,[tool length offset register number]
Example: LOADTL/1,ADJUST
Meaning: Results in tape code to load tool number 1 into the spindle and call
       up tool length offset register number 1.

To change a tool when premeasured tools are used:
LOADTL/[tool no.],LENGTH,[tool length],ADJUST,[tool register]
Example: LOADTL/2,LENGTH,9,ADJUST,2

Meaning: Results in tape code to load tool number 2 into the spindle and call up tool length offset register number 2. All Z-axis coordinates will be modified by the 9-inch length of the tool.

Spindle Speeds.

To define a clockwise spindle speed:
SPINDL/[RPM],CLW
Example: SPINDL/2000,CLW
Meaning: Spindle speed is 2000 RPM in a clockwise direction.

To define a counterclockwise spindle speed:
SPINDL/[RPM],CCLW
Example: SPINDL/2300,CCLW
Meaning: Spindle speed is 2300 RPM in a counterclockwise direction.

Machine Statements. Machine statements tell the computer where to find the instructions to write a program for a particular machine. The syntax is as follows:

MACHIN/[machine identifier]
Example: MACHIN/UNIV 1
Meaning: The machine to program is Universal machine #1.

Tool Statements. Tool statements define to the computer the various tools that will be used during the program. There are two types of cutter definitions: simple and complex. Simple cutters have only a diameter and radius. Complex cutters have multiple features. Figure 15-7 shows both simple and complex cutters. The tool definition is done with a CUTTER statement.

Simple cutter:
CUTTER/d,r
Example: CUTTER/.75,.0625
Meaning: The cutter has a diameter of .750 inch and a corner radius of .0625 inch. The height of the cutter (h) is assumed to be 5.0 inches.

Complex cutter:
CUTTER/d,r,e,f,a,b,h
Example: CUTTER/.75,.0625,.3125,.0837,15,5,6
Meaning: The cutter has a diameter of .750 inch, a corner radius of .0625 inch, E distance is .3125 inch, F distance is .0837 inch, angle A is 15 degrees, angle B is 5 degrees, and the height of the cutter is 6.0 inches.

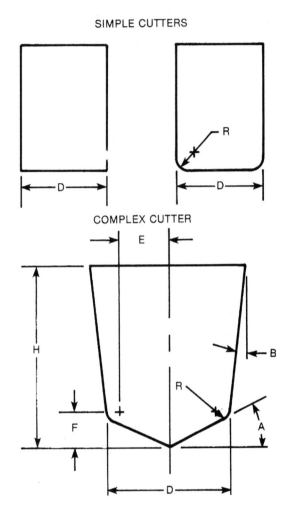

SIMPLE CUTTERS

COMPLEX CUTTER

FIGURE 15-7
Cutter dimensions

Tool Motion Statements in APT

Once the part geometry has been defined, the cutter path can be described to the computer using what are known as tool motion statements. Tool motion is controlled using the following commands: GOUP, GODOWN, GOFWD, GOBACK, GORGT (right), and GOLFT (left). The relationship of these commands can be seen in Figure 15-8.

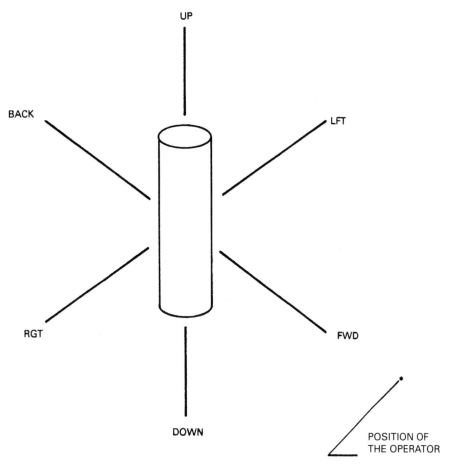

FIGURE 15-8
Tool motion commands in APT

Tool movement is also controlled by three surfaces known as the part surface, the drive surface, and the check surface (see Figure 15-9). The tool bottom is guided by the part surface, while the side of the tool is guided by the drive surface. Once initiated, tool motion continues until stopped by a check surface, which signals the end of the cut.

The six tool commands are used with one of four modifiers to define the check surface. These modifiers—TO, ON, PAST, and TANTO—are illustrated in Figure 15-10.

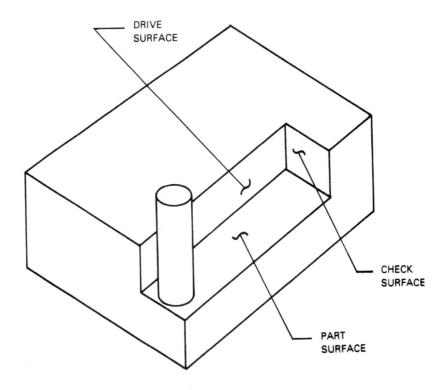

DRIVE
SURFACE

CHECK
SURFACE

PART
SURFACE

FIGURER 15-9
Controlling surfaces in APT

Figure 15-11 shows a rectangular piece defined by the APT language. The four sides have been labeled S1 through S4. Tool motion statements to move the tool from position A to B, and then mill the periphery of the part would be as follows:

GOTO/S1 ,PASTS2
GOFWD/S 1, PASTS4
GORGT/S4, PASTS3
GORGT/S3, PASTS2
GORGT/S2, PASTS1

Figure 15-12 shows a part to be APT programmed. Figure 15-13 identifies the geometric figures used to define the part. Figure 15-14 is a simplified APT program written to mill the part periphery.

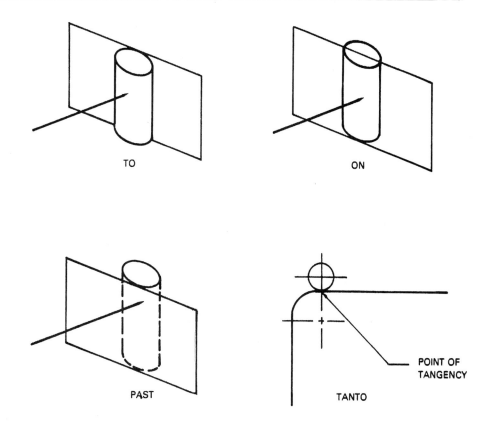

FIGURE 15-10
Modifiers for tool motion

Part Geometry Definitions in COMPACT II

To Define a Point. Points may be defined by Cartesian coordinates or by relationship to other geometric features such as lines or circles.

Point defined by Cartesian coordinates, Figure 15-3(a):
DPTn,nXB,nYB,nZB
Example: DPT1,6XB,6YB,6ZB
Meaning: Point 1 is the point X6.0000, Y6.0000, Z6.0000 from 0/0. The B specifies that a work coordinate system, called a BASE, is being referenced.

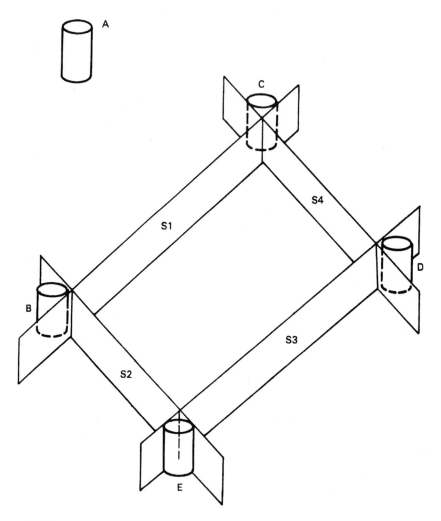

FIGURE 15-11

Point defined by the intersection of two lines, Figure 15-3(b):
DPTn,LNn,LNn
Example: DPT2,LN1,LN2
Meaning: Point 2 is the point formed by the intersection of line 1 and line 2.

Point defined by the center of a circle, Figure 15-3(c):
DPTn,CIRn,CNTR
Example: DPT3,CIR1,CNTR
Meaning: Point 3 is the point at the center of circle 1.

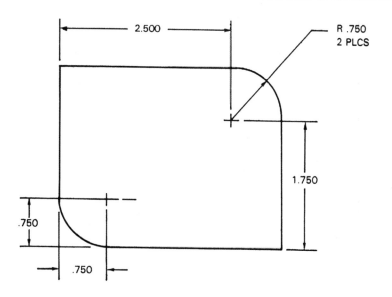

FIGURE 15-12
Part to be APT programmed

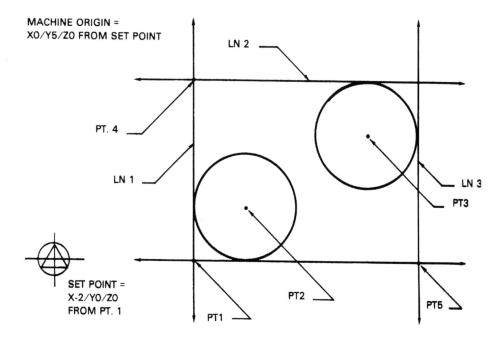

FIGURE 15-13
Geometry of part in Figure 15-12

To Define a Line.

Lines defined by two points, Figure 15-4(a):
DLNn,PTn,PTn
Example: DLN1,PT1,PT2
Meaning: Line 1 is the line passing through point 1 and point 2.

Lines defined by a point and parallel line, Figure 15-4(b):
DLNn,PTn,PARLNn
Example: DLN2,PT2,PARLN1
Meaning: Line 2 is the line passing through point 2 parallel to line 1.

Lines defined by a point and a perpendicular line, Figure 15-4(c):
DLNn,PTn,PERLNn
Example: DLN3,PT3,PERLN2
Meaning: Line 3 is the line passing through point 3 perpendicular to line 2.

Lines defined through a point and tangent to a circle, Figure 15-4(d):
DLNi,PTi,CIRi,MODIFIER
Example: DLN4,PT2,CIR1,YL
Meaning: Line 4 is the line passing through point 2, tangent to circle 1 on the
Y large (right) side.

To Define a Circle.

Circle defined by the centerpoint and radius, Figure 15-5(a):
DCIRn,PTn,R
Example: DCIR1,PT1,1.0R
Meaning: Circle 1 is the circle with center at point 1 and a radius of 1.0 inches.

Circle defined by three points on the circumference, Figure 15-5(d):
DCIRn,PTn,PTn,PTn
Example: DCIR2,PT2,PT3,PT4
Meaning: Circle 2 is the circle with points 2, 3, and 4 on its circumference.

Auxiliary Statements in COMPACT II

Five auxiliary statements are used to initiate and set up the COMPACT II program.

The machine statement MACHIN is used first. MACHIN,UNIVER1 would signal the computer that a machine called UNIVER (Universal) 1 was being used. This tells the computer which postprocessor to use.

The next statement is called the identification statement. This statement identifies the program so that it can be cataloged and retrieved at a later date.

```
MACHIN/UNIVERSAL-1
$$
$$GEOMETRY
$$
PL1      =PLANE/XYPLAN,-1

SETPT    =POINT/-2,0,0
PT1      =POINT/0,0,-1
PT2      =POINT/.75,.75,-1
PT3      =POINT/2.75,1.75,-1
PT4      =POINT/0,2.5,-1
PT5      =POINT/3.25,0,-1

CIR1     =CIRCLE/CENTER,PT2,RADIUS,.75
CIR2     =CIRCLE/CENTER,PT3,RADIUS,.75

LN1      =LINE/PT1,PT4
LN2      =LINE/PT2,LEFT,TANTO,CIR2
LN3      =LINE/PT5,PARLEL,LN1
LN4      =LINE/PT1,PT5
$$
$$MOTION
$$
FROM/SETPT
LOADTL/1,ADJUST,1
CUTTER/.75
SPINDL/2000,CLW
FEDRAT/12,IPM
RAPID,GO/LN1,PL1,PAST,LN4
TLLFT,GOLFT/LN1,PAST,LN2
GORGT/LN2,TANTO,CIR2
GOFWD/CIR2,TANTO,LN3
GOFWD/LN3,PAST,LN4
GORGT/LN4,PAST,LN1
RAPID,GOTO/SP
FINI
```

FIGURE 15-14
APT program to mill the part in Figure 15-12

The word IDENT begins the statement. IDENT,EXAMPLE,PARTNO.1 would identify the program as example 1, part number 1.

The next statement is called the initialization statement (INIT). This statement tells the computer what modes will be used for input and output. INIT,INCH/IN,INCH/OUT tells the computer that all input and output dimensions are in inches.

The fourth statement in the program is the setup statement (SETUP). This statement identifies the machine origin, tool change location, tool travel limits, positioning system to be used, and any other special requirements that may be necessary for the particular machine.

The fifth statement is the base statement. The base statement establishes a work coordinate system. BASE 6XA,6YA,6ZA establishes the work coordinate system 6.0 inches in X, 6.0 inches in Y, and 6.0 inches in Z from the machine origin. The A following the axes indicates machine absolute system (the machine origin). In geometry definition statements, the suffix B is used to indicate base coordinate system (the work coordinate system).

Other statements used in COMPACT II are: ATCHG to initiate an automatic tool change; CON to turn the coolant on; STK to identify the amount of stock to be left for finishing; FRM to assign a feedrate in inches per minute; IPR to assign a feedrate in inches per revolution.

Tool Motion Statements in COMPACT II

The tool motion statements used in COMPACT II are similar to those used in APT. The modifiers ON, TO, and PAST are used in identical fashion. Other statements are:

MOVE—To designate a move in rapid traverse.
CUT—To designate a move at feedrate.
ICON—To identify an inside contour.
OCON—To identify an outside contour.
S—To identify the starting location of a cut.
F—To identify the finish location of a cut.

In Figure 15-15 the geometry of a part is illustrated. The COMPACT II statements to move the tool from home to #1 and then around the part periphery are:

MOVE,TOLN 1, PASTLN4
CUT, PARLN 1, PASTLN2
CUT,PARLN2,TAN,CIR1
OCON,CIR1,CW,S(90),F(0)

S(90) means the starting point is at 90 degrees; F(0) means the finish point is at 0 degrees.

The Postprocessor

The postprocessor is a separate program that translates the part program into the codes necessary for the particular machine defined to the computer through the machine statement. *Postprocessor is* really a short term for postprocessing program. Most machinery programmed using computer-aided programming languages uses RS-274 (word address) format. The postprocessor converts the APT or COMPACT II program (part geometry, auxiliary, and tool motion statements) to a word address format numerical control program. This program may then be punched into a tape using either

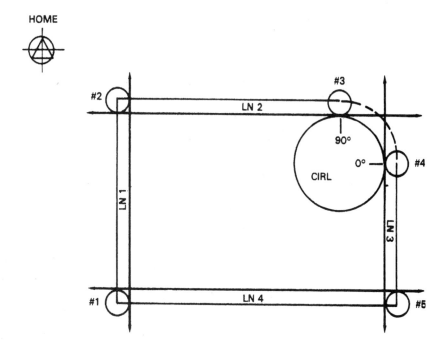

HOME

FIGURE 15-15

RS-244 or RS-358 coding; or it may be transferred directly from the main computer to the machine's computer. The advantage of computer-aided programming languages over straight word address programming is the computer's ability to perform the calculations for very complex surfaces. Manual programming of parts requiring four and five axes is far more complicated than two- and three-axis programming. Figure 15-16 shows a part for which an APT program is written and postprocessed. Figure 15-17 shows the part geometry, and Figure 15-18 shows the APT program and the postprocessor output.

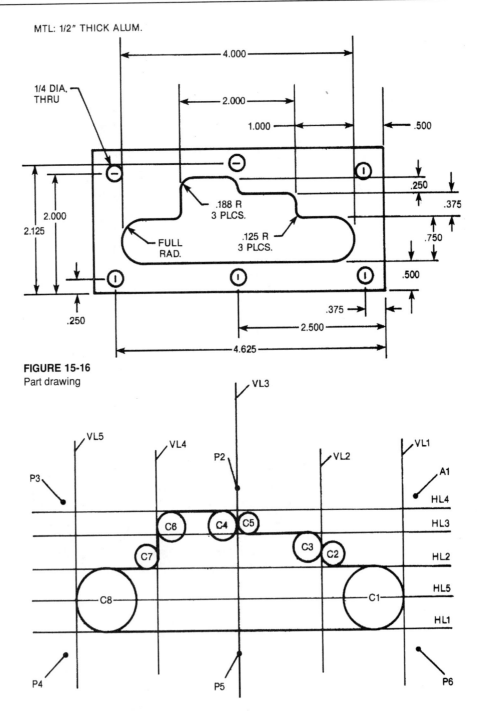

MTL: 1/2" THICK ALUM.

1/4 DIA.
THRU

4.000

2.000

1.000   .500

.188 R
3 PLCS.

.250

.375

FULL
RAD.

.125 R
3 PLCS.

.750

2.000

2.125

.500

.250

.375

2.500

4.625

FIGURE 15-16
Part drawing

VL3

VL5   VL4   P2   VL2   VL1

P3   A1

                                                             HL4

                        C6   C4   C5   HL3

                                                C3   HL2
                   C7   C2

                                                             HL5
        C8   C1
                                                             HL1

P4   P5   P6

FIGURE 15-17
Part geometry

```
0001 PARTNO APT MILL EXAMPLE
0002 MACHIN/UNCM01,01
0003 MACHIN/PCPLOT,1
0004 $$
0005 $$MACRO TO PRINT A ROW OF ASTERIX ACROSS PAGE
0006 ASTRX  =MACRO/
0007 PPRINT ****************************************************
0008 TERMAC
0009 $$
0010 CALL/ASTRX
0011 PPRINT COORDINATE SYSTEM ORIGIN:
0012 PPRINT XO = LOWER RIGHT CORNER OF PART
0013 PPRINT YO = LOWER RIGHT CORNER OF PART
0014 PPRINT ZO = 2.000 INCHES ABOVE TOP OF PART
0015 CALL/ASTRX
0016 PPRINT TOOL LIST
0017 PPRINT TOOL 1:   3/8 DIA. 2 FLUTE CARBIDE END MILL
0018 PPRINT TOOL 2:   1/4 DIA. 2 FLUTE CARBIDE END MILL
0019 PPRINT TOOL 3:   NO. 4 X 90 DEG. C'DRILL
0020 PPRINT TOOL 4:   1/4 DRILL
0021 CALL/ASTRX
0022 PPRINT MACHINING SEQUENCE
0023 PPRINT TOOL 1:   SPEED 3800 RPM, FEEDRATE .003 IPR, CRO REGISTER D11
0024 PPRINT            ROUGH MILL INDSIDE CONTOUR OF PART.
0025 PPRINT
0026 PPRINT TOOL 2:   SPEED 4000 RPM, FEEDRATE .002 IPR, CRO REGISTER D21
0027 PPRINT            FINISH MILL INSIDE CONTOUR OF PART.
0028 PPRINT
0029 PPRINT TOOL 3:   SPEED 3500 RPM, FEEDRATE .004 IPR
0030 PPRINT            C'DRILL & C'SINL 1/4 HOLES TO .260 DIA. 6 PLCS.
0031 PPRINT
0032 PPRINT TOOL 4:   SPEED 3000 RPM, FEEDRATE .004 IPR
0033 PPRINT            DRILL 1/4 DIA. THRU 6 PLCS.
0034 CALL/ASTRX
0035 $$
0036 $$PART GEOMETRY
0037 $$
0038 PL1   =PLANE/XYPLAN,-2                  $$TOP OF PART
0039 PL2   =PLANE/PARLEL,PL1,ZSMALL,.51      $$.01 BELOW BOTTOM OF PART
0040 CLR   =PLANE/PARLEL,PL1,ZLARGE,.05      $$.05 ABOVE TOP OF PART
0041 VL1   =LINE/YAXIS,-.5
0042 VL2   =LINE/PARLEL,VL1,XSMALL,1
0043 VL3   =LINE/YAXIS,-2.5
0044 VL4   =LINE/PARLEL,VL2,XSMALL,2
0045 VL5   =LINE/PARLEL,VL1,XSMALL,4

0046 HL1   =LINE/XAXIS,.5
0047 HL2   =LINE/PARLEL,HL1,YLARGE,.75
```

FIGURE 15-18 *(Continues to page 354)*

```
0048 HL3   =LINE/PARLEL,HL2,YLARGE,.375
0049 HL4   =LINE/PARLEL,HL3,YLARGE,.25
0050 HL5   =LINE/PARLEL,HL1,YLARGE,.375

0051 C1    =CIRCLE/XSMALL,VL1,YLARGE,HL1,YSMALL,HL2
0052 C2    =CIRCLE/XLARGE,VL2,YLARGE,HL2,RADIUS,.125
0053 C3    =CIRCLE/XSMALL,VL2,YSMALL,HL3,RADIUS,.188
0054 C4    =CIRCLE/XSMALL,VL3,YSMALL,HL4,RADIUS,.188
0055 C5    =CIRCLE/YLARGE,HL3,XLARGE,OUT,C4,RADIUS,.125
0056 C6    =CIRCLE/XLARGE,VL4,YSMALL,HL4,RADIUS,.188
0057 C7    =CIRCLE/XSMALL,VL4,YLARGE,HL2,RADIUS,.125
0058 C8    =CIRCLE/YLARGE,HL1,XLARGE,VL5,YSMALL,HL2

0059 ZSURF/PL1
0060      P1    =POINT/-.375,2
0061      P2    =POINT/-2.5,2.125
0062      P3    =POINT/-4.625,2
0063      P4    =POINT/-4.625,.25
0064      P5    =POINT/-2.5,.25
0065      P6    =POINT/-.375,.25
0066      PAT1 =PATERN/P1,P2,P3,P4,P5,P6
0067 $$
0068 $$MOTION STATEMENTS
0069 $$
0070 PPLOT/1
0071 FROM/0,4,0
0072 CALL/ASTRX
0073 PPRINT 3/8 END MILL - USES CRO REGISTER D11
0074 PPRINT ROUGH MILL INSIDE CONTOUR - LEAVE .01 STK. TO FINISH.
0075 LOADTL/1,ADJUST,1
0076 CUTTER/.375
0077 SPINDL/3800,CLW
0078 COOLNT/ON
0079 FEDRAT/.003,IPR

0080 PPRINT
0081 PPRINT POSITION TO START XY AND FEED TO DEPTH
0082 RAPID,GO/ON,VL2,CLR,ON,HL5
0083 THICK/0,.01,.01
0084 GO/ON,VL2,PL2
0085 CUTCOM/LEFT,11

0086 PPRINT
0087 PPRINT MILL .750 WIDE SLOT
0088 GO/HL2
0089      GOLFT/HL2,TANTO,C8
0090      GOFWD/C8,TANTO,HL1
0091      GOFWD/HL1,TANTO,C1
0092      GOFWD/C1,TANTO,HL2
0093      GOFWD/HL2,PAST,VL2

0094 PPRINT
0095 PPRINT MILL BALANCE OF SLOT
0096 GO/HL3
```

```
0097      GOLFT/HL3,PAST,VL3
0098      GORGT/VL3,HL4
0099      GOLFT/HL4,VL4
0100      GOLFT/VL4,ON,HL5

0101 PPRINT
0102 PPRINT RETRACT SPINDLE, CANCEL CUTTER COMP, CANCEL TOOL OFFSET
0103 RETRCT
0104 CUTCOM/OFF
0105 COOLNT/OFF
0106 RESET
0107 $$
0108 PPLOT/2
0109 CALL/ASTRX
0110 PPRINT 1/4 END MILL - USES CRO REGISTER D21
0111 PPRINT FINISH MILL SLOT
0112 LOADTL/2,ADJUST,2
0113 CUTTER/.250
0114 SPINDL/4000,CLW
0115 COOLNT/ON
0116 FEDRAT/.002,IPR

0117 PPRINT
0118 PPRINT POSTITON TO CENTER OF C1 AND FEED TO Z DRPTH
0119 THICK/0
0120 RAPID,GO/ON,(LINE/PARLEL,VL1,XSMALL,.375),CLR,ON,HL5
0121 NOPS,GO/PL2
0122 AUTOPS

0123 PPRINT
0124 PPRINT FINISH MILL CONTOUR OF I.D.
0125 CUTCOM/LEFT,21
0126 GO/HL2
0127      GOLFT/HL2,TANTO,C2
0128      GOFWD/C2,TANTO,VL2
0129      GOFWD/VL2,TANTO,C3
0130      GOFWD/C3,TANTO,HL3
0131      GOFWD/HL3,TANTO,C5
0132      GOFWD/C5,TANTO,C4
0133      GOFWD/C4,TANTO,HL4
0134      GOFWD/HL4,TANTO,C6
0135      GOFWD/C6,TANTO,VL4
0136      GOFWD/VL4,TANTO,C7
0137      GOFWD/C7,TANTO,HL2
0138      GOFWD/HL2,TANTO,C8
0139      GOFWD/C8,TANTO,HL1
0140      GOFWD/HL1,TANTO,C1
0141      GOFWD/C1,TANTO,HL2
0142      GOFWD/HL2,PAST,VL2

0143 PPRINT
0144 PPRINT RETRACT SPINDLE, CANCEL CUTTER COMP, CANCEL TOOL OFFSET
0145 RETRCT
0146 CUTCOM/OFF
```

```
0147 COOLNT/OFF
0148 RESET
0149 $$
0150 PPLOT/3
0151 CALL/ASTRX
0152 PPRINT NO. 4 C'DRILL - C'DRILL 6 PLCS. TO .260 DIA.
0153 LOADTL/3,ADJUST,3
0154 SPINDL/3500,CLW
0155 COOLNT/ON
0156 PREFUN/99,NEXT
0157 CYCLE/DRILL,.183,.004,IPR,.1
0158      GOTO/PAT1
0159 CYCLE/OFF
0160 COOLNT/OFF
0161 RESET
0162 $$
0163 DRAFT/OFF
0164 CALL/ASTRX
0165 PPRINT 1/4 DRILL - PECK DRILL .250 THRU 6 PLCS.
0166 LOADTL/4,ADJUST,4
0167 SPINDL/3000,CLW
0168 COOLNT/ON
0169 PREFUN/99,NEXT
0170 CYCLE/DEEP,.6,.003,IPR,.1,INCR,.25
0171      GOTO/PAT1
0172 CYCLE/OFF
0173 COOLNT/OFF
0174 RESET
0175 FINI
```

NO DIAGNOSTICS ELICITED DURING TRANSLATION PHASE
185 N/C SOURCE RECORDS (SYSIN)

SECTION 1 ELAPSED CPU TIME IN MIN/SEC IS 0000/00.4433
SECTION 2 ELAPSED CPU TIME IN MIN/SEC IS 0000/00.1866

```
UNCM01 03.000.I370 MACHIN/UNCM01, 1   DATE 88.319    PAGE   1
APT MILL EXAMPLE                                                    (INCH)
INPUT  CLREC N3G2X34Y34R34Z34Q34I34J34K34A43P4F32S4T2D2H2M2
    7      9  APT MILL EXAMPLE
    7      9  $
    7      9  LEADER/   20.0
    7      9  $
    7      9  N001 G90 G00 G70 G80 G40$
    7      9  ***************************************************
   11     11  COORDINATE SYSTEM ORIGIN:
   12     13  X0 = LOWER RIGHT CORNER OF PART
   13     15  Y0 = LOWER RIGHT CORNER OF PART
   14     17  Z0 = 2.000 INCHES ABOVE TOP OF PART
    7     20  ***************************************************
   16     22  TOOL LIST
   17     24  TOOL 1:  3/8 DIA. 2 FLUTE CARBIDE END MILL
   18     26  TOOL 2:  1/4 DIA. 2 FLUTE CARBIDE END MILL
   19     28  TOOL 3:  NO. 4 X 90 DEG. C'DRILL
   20     30  TOOL 4:  1/4 DRILL
    7     33  ***************************************************
   22     35  MACHINING SEQUENCE
   23     37  TOOL 1:  SPEED 3800 RPM, FEEDRATE .003 IPR, CRO REGISTER D11
   24     39           ROUGH MILL INDSIDE CONTOUR OF PART.
   25     41
   26     43  TOOL 2:  SPEED 4000 RPM, FEEDRATE .002 IPR, CRO REGISTER D21
   27     45           FINISH MILL INSIDE CONTOUR OF PART.
   28     47
   29     49  TOOL 3:  SPEED 3500 RPM, FEEDRATE .004 IPR
   30     51           C'DRILL & C'SINL 1/4 HOLES TO .260 DIA. 6 PLCS.
   31     53
   32     55  TOOL 4:  SPEED 3000 RPM, FEEDRATE .004 IPR
   33     57           DRILL 1/4 DIA. THRU 6 PLCS.
    7     60  ***************************************************
   71     64  FROM/XYZ =     0.0000    4.0000    0.0000
    7     67  ***************************************************
   73     69  3/8 END MILL - USES CRO REGISTER D11
   74     71  ROUGH MILL INSIDE CONTOUR - LEAVE .01 STK. TO FINISH.
   75     73  TOOL TIME =    0.04
   75     73  $
   75     73  N002 G90 G00 X.0000 Y4.0000 T01$
   75     73  N003 S0500 M03$
   75     73  N004 G44 Z.0000 H01$
   77     77  N005 S3800$
   78     79  N006 M08$
   80     83
   81     85   POSITION TO START XY AND FEED TO DEPTH
   82     88  N007 G00 X-1.5000 Y.8750$
   82     88  N008 Z-1.9500$
                        TAPE    TIME    WARNING
                PAGE    6.38    0.15       0
                TOTAL   6.38    0.15       0
```

```
UNCM01 03.000.I370 MACHIN/UNCM01, 1   DATE 88.319    PAGE   2
APT MILL EXAMPLE                                                  (INCH)
INPUT  CLREC N3G2X34Y34R34Z34Q34I34J34K34A43P4F32S4T2D2H2M2
    84      92 N009 G01 Z-2.5100 F11.40$
    85      94 N010 G17$
    86      96
    87      98   MILL .750 WIDE SLOT
    88     100 N011 G41 X-1.5000 Y.8760 J1.0000 D11$
    88     100 N012 G01 Y1.0525$
    89     102 N013 X-4.1250$
    91     107 N014 G03 X-4.1250 Y.6975 I-4.1250 J.8750$
    91     107 N015 G01 X-.8750$
    93     112 N016 G03 X-.8750 Y1.0525 I-.8750 J.8750$
    93     112 N017 G01 X-1.6975$
    94     114
    95     116   MILL BALANCE OF SLOT
    96     118 N018 Y1.4275$
    97     120 N019 X-2.6975$
    98     122 N020 Y1.6775$
    99     124 N021 X-3.3025$
   100     126 N022 Y.8750$
   101     128
   102     130   RETRACT SPINDLE, CANCEL CUTTER COMP, CANCEL TOOL OFFSET
   103     132 N023 G00 Z.0000$
   104     134 N024 G40$
   105     136 N025 M09$
   106     138 N026 G00 Z.0000$
   106     138 N027 G49$
   106     138 N028 M01$
     7     143 *********************************************************
   110     145   1/4 END MILL - USES CRO REGISTER D21
   111     147   FINISH MILL SLOT
   112     149 TOOL TIME =    1.09
   112     149 $
   112     149 N029 G90 G00 X-3.3025 Y.8750 T02$
   112     149 N030 S0500 M03$
   112     149 N031 G44 Z.0000 H02$
   114     153 N032 S4000$
   115     155 N033 M08$
   117     159
   118     161   POSTITON TO CENTER OF C1 AND FEED TO Z DRPTH
   120     166 N034 G00 X-.8750$
   120     166 N035 Z-1.9500$
   121     168 N036 G01 Z-2.5100 F8.00$
   123     171
   124     173   FINISH MILL CONTOUR OF I.D.
   126     177 N037 G41 X-.8750 Y.8760 J1.0000 D21$
   126     177 N038 G01 Y1.1250$
```

	TAPE	TIME	WARNING
PAGE	3.70	1.21	0
TOTAL	10.08	1.36	0

```
UNCM01 03.000.I370 MACHIN/UNCM01, 1    DATE 88.319      PAGE   3
APT MILL EXAMPLE                                                    (INCH)
INPUT  CLREC N3G2X34Y34R34Z34Q34I34J34K34A43P4F32S4T2D2H2M2
  127     179 N039 X-1.3750$
  129     184 N040 G02 X-1.6250 Y1.3750 I-1.3750 J1.3750$
  129     184 N041 G01 Y1.4370$
  131     189 N042 G03 X-1.6880 Y1.5000 I-1.6880 J1.4370$
  131     189 N043 G01 X-2.3814$
  133     194 N044 G02 X-2.6263 Y1.6997 I-2.3814 J1.7500$
  134     197 N045 G03 X-2.6880 Y1.7500 I-2.6880 J1.6870$
  134     197 N046 G01 X-3.3120$
  136     202 N047 G03 X-3.3750 Y1.6870 I-3.3120 J1.6870$
  136     202 N048 G01 Y1.3750$
  136     207 N049 G02 X-3.6250 Y1.1250 I-3.6250 J1.3750$
  138     207 N050 G01 X-4.1250$
  140     212 N051 G03 X-4.1250 Y.6250 I-4.1250 J.8750$
  140     212 N052 G01 X-.8750$
  142     217 N053 G03 X-.8750 Y1.1250 I-.8750 J.8750$
  142     217 N054 G01 X-1.6250$
  143     219
  144     221   RETRACT SPINDLE, CANCEL CUTTER COMP, CANCEL TOOL OFFSET
  145     223 N055 G00 Z.0000$
  146     225 N056 G40$
  147     227 N057 M09$
  148     229 N058 G00 Z.0000$
  148     229 N059 G49$
  148     229 N060 M01$
    7     234 *********************************************************
  152     236   NO. 4 C'DRILL - C'DRILL 6 PLCS. TO .260 DIA.
  153     238 TOOL TIME =     1.38
  153     238 $
  153     238 N061 G90 G00 X-1.6250 Y1.1250 T03$
  153     238 N062 S0500 M03$
  153     238 N063 G44 Z.0000 H03$
  154     240 N064 S3500$
  155     242 N065 M08$
  158     248 N066 G81 G99 X-.3750 Y2.0000 R-1.9000 Z-2.1830 F14.00$
  158     249 N067 X-2.5000 Y2.1250$
  158     250 N068 X-4.6250 Y2.0000$
  158     251 N069 Y.2500$
  158     252 N070 X-2.5000$
  158     253 N071 X-.3750$
  159     255 N072 G80$
  160     257 N073 M09$
  161     259 N074 G00 Z.0000$
  161     259 N075 G49$
  161     259 N076 M01$
    7     264 *********************************************************
                   TAPE      TIME    WARNING
            PAGE    5.63     1.37        0
            TOTAL  15.72     2.73        0
```

UNCM01 03.000.I370 MACHIN/UNCM01, 1   DATE 88.319   PAGE   4
APT MILL EXAMPLE   (INCH)
INPUT   CLREC N3G2X34Y34R34Z34Q34I34J34K34A43P4F32S4T2D2H2M2
   165   266 1/4 DRILL - PECK DRILL .250 THRU 6 PLCS.
   166   268 TOOL TIME =   0.21
   166   268 $
   166   268 N077 G90 G00 X-.3750 Y.2500 T04$
   166   268 N078 S0500 M03$
   166   268 N079 G44 Z.0000 H04$
   167   270 N080 S3000$
   168   272 N081 M08$
   171   278 N082 G83 G99 X-.3750 Y2.0000 R-1.9000 Z-2.6000 Q.2500 F9.00$
   171   279 N083 X-2.5000 Y2.1250$
   171   280 N084 X-4.6250 Y2.0000$
   171   281 N085 Y.2500$
   171   282 N086 X-2.5000$
   171   283 N087 X-.3750$
   172   285 N088 G80$
   173   287 N089 M09$
   174   289 N090 G00 Z.0000$
   174   289 N091 G49$
   174   289 N092 M01$
   175   291 TOOL TIME =   0.56
   175   291 N093 G00 Y4.0000 T15$
   175   291 N094 M30$
   175   291 =$
                      TAPE   TIME   WARNING
              PAGE   2.13   0.56   0
              TOTAL 17.85   3.29   0

UNCM01 03.000.I370 MACHIN/UNCM01, 1   DATE 88.319   PAGE   5
APT MILL EXAMPLE   (INCH)
INPUT   CLREC N3G2X34Y34R34Z34Q34I34J34K34A43P4F32S4T2D2H2M2
              TOOL SEQUENCE

INPUT	CLREC	TOOL	OFFSET	LENGTH Z
75	73	1	1	0.0000
112	149	2	2	0.0000
153	238	3	3	0.0000
166	268	4	4	0.0000

       TOTAL PART PROGRAM CPU TIME IN MIN/SEC IS 0000/02.3966
              **** END OF APT PROCESSING ****

FIGURE 15-18
APT program and postprocessor output for part shown in Figure 15-16

COMPUTER GRAPHICS PROGRAMMING

The newest form of NC programming is called CAM. CAM stands for computer-aided manufacturing. When using a CAM system, the programmer either calls up an existing part drawing, or defines the part geometry to the computer. Next, the cutter path is drawn around the part and the necessary information on cut direction, tool, speeds, and feeds are input. This information is then converted into either an APT file, or a cutter centerline data file by a postprocessor. This data is then fed through a secondary postprocessor, which produces the necessary NC code for a given machine.

There are a number of different CAM systems on the market. The following is a brief explanation of several different types.

Digitizing Systems

Digitizing systems use an existing part drawing to obtain the geometry information. The drawing used must be drawn to a true scale of the finished part. The scale drawing is fed into a digitizer which is connected to the CAM system computer. A digitizer is a device consisting of a table with a probe or other sensor attached. The sensing device is passed over the drawing, converting drawing lines in the necessary mathematical information into electronic form which the computer needs to recreate the drawing. The cutter path is then defined by the programmer and the cutter, speed, and feed information input. The result is then postprocessed into the necessary tape code for the CNC machine.

Digitizing is one of the simplest CAM methods, but it is also the least accurate. The accuracy of the part geometry is dependent upon the accuracy of the scaled drawing which was digitized. Fortunately this shortcoming is minimized by the fact that drawings 30 times size or larger can be used.

Scanning Systems

Scanning systems use the part itself rather than a part drawing to obtain the geometric database to machine the part. Scanning is used when complex curves are to be machined which are difficult to draw and which do not fit a true mathematical model. Automobile bodies are an example of such curves. A scanner is a probing device connected to the CAM system computer which is passed over a model of the part. The probe feeds information concerning the part geometry into the computer which then calculates the points necessary to define the part shape. The cutter path, tool data, speeds, and feeds are then input. The information is then fed to the postprocessor which will convert the information into the necessary tape coding.

CAM Systems

Although the two methods mentioned previously are referred to as CAM systems, CAM is also the term used to describe computer graphics programming. CAM systems can be run on a mainframe computer, or on a PC. Mainframe systems are found in larger plants primarily involved with four- or five-axis programming. Microcomputer-based CAM systems function well for three-axis. New systems can handle four-axis programming as well. Microcomputer systems are an economical choice for many small to midsized shops. Figure 15-19 illustrates a PC-based CAM system running on an APPLE Macintosh. Figure 15-20 depicts another microbased system.

In CAM (graphics) programming, the programmer defines the part geometry to the computer using one or more of several input devices. These devices may be the keyboard, a mouse, a digitizer, or a light pen. After the part geometry is defined, the cutter path is drawn around the part. Information on the cut direction, tool data, speeds, and feeds are then input. This information is then translated by a series of postprocessors into the necessary NC code. Figure 15-21 shows a cutter and tool path generated on a mainframe-based CAM system.

FIGURE 15-19
Mac EZ-CAM II *(Photo courtesy of Bridgeport Machine Inc.)*

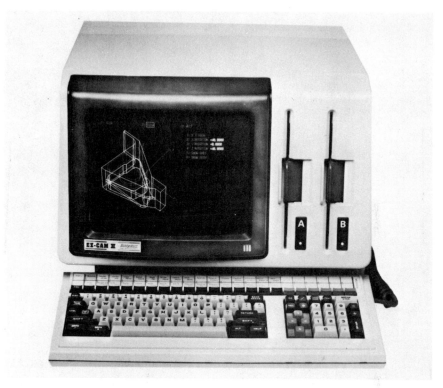

FIGURE 15-20
EZ-CAM II *(Photo courtesy of Bridgeport Machine Inc.)*

Once programmed and postprocessed, the cutter path can be plotted out on paper as a preliminary prove-out of the program. Figure 15-22 shows one type of plotter. The program plot can help the programmer spot errors that have been overlooked when the cutter path was on the computer monitor screen.

There are many types of CAM systems on the market. One need only glance at the pages of an industry trade magazine to appreciate the number that exist.

CAD/CAM Programming

CAD/CAM programming is one of the more sophisticated CAM programming systems. In a CAD/CAM system, the part geometry is obtained from the CAD (computer-aided design) drawing itself. The cutter path, tool, speeds, and feeds are then defined and the program postprocessed into the NC tape code. There are two basic types of CAD/CAM systems: stand alone and modular.

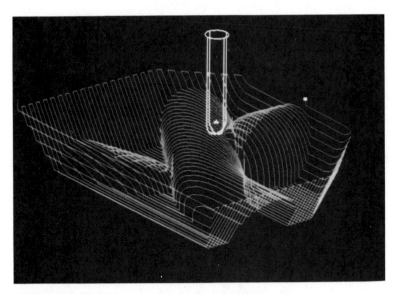

FIGURE 15-21
Cutter and tool path generated on a mainframe-based CAM system *(Photo courtesy of Battelle Inc.)*

Stand alone systems contain both the CAD and the CAM modules in one system, furnished by one supplier. Generally, these systems are intended to function only with the supplied modules and will not link with other vendors products.

Modular systems consist of a CAD package made by one manufacturer, and a CAM package by another. The CAM packages are designed to link to the most popular CAD packages such as AutoCad, Micro CADAM, and GEOSPOT.

CAD/CAM Programming

Referring back to Figure 15-2, there are three components of a CAD/CAM program: part geometry definition, tool motion commands, and auxiliary statements. The first part of writing any CAM program is defining the part to the computer. This can be done either by the CAM programs internal CAD drawing package (if so equipped), or using another CAD package. If the part was originally drawn using a CAD system, that CAD drawing file can be used as the basis for the machining program. Invariably the programmer will need to modify the geometry at some point to meet his/her needs. Geometry definition is accomplished using a series of menu selected options and commands. Input of the information is done using both the keyboard and a pointing device (a mouse or digitizer pad). The geometry is then saved in a separate geometry file. This differs from language-based programming where the geometry is internal to the source code file.

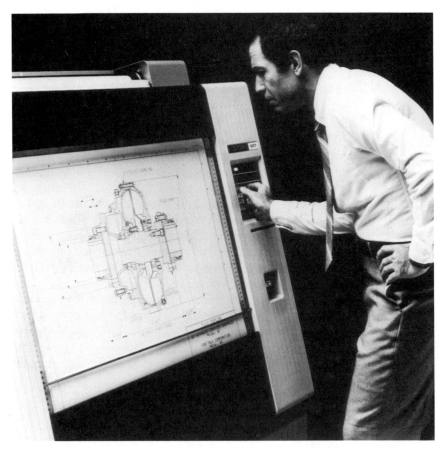

FIGURE 15-22
Plotter *(Photo courtesy of Calccomp Inc.)*

After the geometry is defined, the programmer describes the tool path for the computer. Generally, the tool path, like the geometry definition, is done using menu-selected options built into the CAM system. Auxiliary commands (such as turning on the coolant, turning on the spindle, etc.) are added to the tool motion section as required. Each tool's motion sequence is defined separately, and stored as an independent file. This allows the programmer to add, change, or delete tools without having to rewrite the entire tool motion.

As a final step before processing, the programmer defines the tool sequence to the computer. The tool sequence file simply contains a list of the tool motion files in the order in which they are to run (i.e., tool #1, tool #2, tool #3, etc.).

When processing the job, the computer takes the information contained in the geometry and tool motion files and generates a CL data file (centerline data file). Once the CL file is generated, the programmer can replay the tool motion. The tool path is graphically simulated on the computer monitor.

Errors detected during replay are corrected in the tool motion file, the CL data file regenerated with the corrections, and the tool motion replayed. After the programmer is satisfied with the results, the CL file is then submitted to the postprocessing program module where it is conveted into an NC code. The NC code file is called the tape image file.

CIM

CAD/CAM programming coupled with DNC systems for program delivery to the individual machines form the fundamental building block for CIM (computer integrated manufacturing). CIM is not a true reality at the time of this publication, but is the wave of the future. In a CIM system, the entire manufacturing process is done with the aid of computers. Product design, manufacturing engineering, production control, quality assurance, procurement, accounting, and management functions would all be linked together in a shared database. The goal of all this is to eliminate paperwork, eliminate duplication of effort, and reduce overall costs while improving part quality and delivery schedules. Much is needed in the way of standardization to achieve a CIM environment, but much has already been accomplished.

For a company to be competitive in the coming years, a move toward computer integration will become a necessity. It is important therefore, for students desiring to make a career in the manufacturing fields to learn and improve upon their computer skills.

SUMMARY

The important concepts presented in this chapter are:

- Computers are used three basic ways in numerical control programming: in offline programming, computer-aided programming, and computer graphics programming.
- Offline programming terminals allow the NC program to be written away from the machine. They do not assist the programmer in any way other than eliminating the need to enter the program at the machine's computer console.
- Computer-aided programming permits the programming of parts for many machines using one programming language. The computer handles the necessary mathematical calculations for the cutter path. A postprocessor then translates this information into codes for a particular machine.
- Three types of statements are used in a computer-aided program: part geometry statements, auxiliary statements, and tool motion statements.
- There are a number of computer-aided programming languages. The one a company uses depends upon the parts it produces and the computers available.

- Graphics programming called CAM programming is being used to simplify the part programming process.
- There are four basic types of CAM systems: digitizing, scanning, CAM, and CAD/CAM.
- There are two basic types of CAD/CAM systems: stand alone and modular.

VOCABULARY INTRODUCED IN THIS CHAPTER

Automatic programmed tools (APT)
Centerline data file (CL file)
Compact II
Computer-aided design (CAD)
Computer-aided manufacturing (CAM)
Computer-aided programming language
Computer-assisted programming
Computer graphics programming
Computer integrated manufacturing (CIM)
Digitizing
Offline programming
Offline programming terminal
Plotter
Postprocessing
Postprocessor
Scanning

REVIEW
QUESTIONS

1. What three ways are computers used in numerical control programming?
2. What is offline programming?
3. What does computer-aided programming allow the programmer to do?
4. What three types of statements are used in computer-aided programs?
5. What is CAM programming?
6. What are four types of CAM systems?
7. What is the difference between a scanning and a digitizing system?
8. What is the difference between a CAM and a CAD/CAM system?

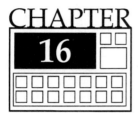

CHAPTER 16

The Future of Numerical Control

OBJECTIVES Upon completion of this chapter, you will be able to:

- Explain why the use of CNC will increase in prototype and small lot job shops.
- Describe a flexible machining system.
- Describe a machining cell.
- Describe the responsibilities of the NC electronics technician, machine operator/ setup operator, and part programmer.

Numerical control will play an increasingly important role in manufacturing in the coming years. CNC is already being applied to machine tools, punch presses, sheet metal brakes, electrical discharge machines, welding machinery, and inspection equipment. Smaller-sized production shops, prototype operations, and large manufacturing concerns will all benefit from recent and continuing developments in numerical control, robotics, and computer technology.

NC IN PROTOTYPE AND JOB SHOPS

The ongoing development of less expensive numerical control systems will offer increasing options to companies that today cannot justify a numerical control system. The lower cost of acquiring machining and turning centers, coupled with the ease of programming and other features of the newest generation of CNC controllers, will result in the adoption of CNC machinery by more and more small job shops. Competition from foreign sources is forcing all companies to look for ways to improve quality while making the changes in design that market conditions so often require. CNC machinery can fulfill both requirements: (1) The repeatability of CNC can improve the overall quality of parts produced; and (2) since CNC uses software programs to produce part shapes, what would have been major retooling becomes the editing and revising of the part program.

Recently, a new type of DNC retrofit system was introduced specifically geared to prototype shop requirements. In this system a microcomputer such as the popular Apple II is used as the controller. The executive program for the system is software, not firmware, requiring little or no modification of the computer. This system is a surprisingly low cost way for a shop to acquire a DNC system. It is not designed for the demands of a manufacturing operation but will handle the one-of-a-kind parts made in prototype, die, or moldmaking shops. Designed for used on vertical mills, it is a three-axis contouring system capable of circular, helical, and linear interpolation.

A problem common to all companies is the shortage of skilled machinists. In smaller companies, the shortage of general machinists, tool and die makers, and mold makers is most acute. In coming years the shortage of skilled prototype machinists and instrument makers is likely to be felt by scientific and research organizations that have their own prototype shops. In addition, increasingly complex part geometries are being required for new technology applications. CNC offers solutions to all these problems.

CNC IN MANUFACTURING

The most exciting developments in NC applications are taking place in large scale manufacturing. In many industries, computer integration of the entire manufacturing process is believed to be possible in the coming decades. The computer capability for *computer integrated manufacturing (C/M)* already exists, but software bases and computer standards to allow networking of design, manufacturing, purchasing, inventory, and marketing functions must be developed and refined. In a CIM system, these various functions are interconnected, using the instant access to information that the computer allows, to eliminate duplication of effort, reduce inventories, reduce part handling, and provide a higher percentage of chip-making time. Although not yet a reality, one of the major building blocks in a CIM system is currently being produced and used by some industries—the flexible machining system (FMS).

Flexible Machining System

A flexible machining system (FMS) is a system of CNC machines, robots, and part transfer vehicles that can take a part from raw stock or casting and perform all necessary machining, part handling, and inspection operations to make a finished part or assembly. It is an entire unmanned, software-based, manufacturing/assembly line. *An* FMS consists of four major components: the CNC machines, coordinate measuring machines, part handling and assembly robots, and part/tool transfer vehicles. Figure 16-1 illustrates a small flexible machining system. This system employs a turning

FIGURE 16-1
(Photo courtesy of Cincinnati Milacron)

center, a horizontal machining center, and a vertical machining center. A single track-guided robot is used as both a load/unload robot and a transfer vehicle.

The main element in an FMS is the CNC machining or turning center. The automatic tool changing capability of these machines allows them to run untended, given the proper support system. Tool monitoring systems built into the CNC machine are used to detect and replace worn tools. The major obstacles in an FMS are not the machining centers but the support systems for the machines, such as part load/unload and part transfer.

Inspection in an FMS is accomplished through the use of coordinate measuring machines. These operate much like CNC machinery in that they are programmed to move to different positions on a workpiece. Instead of using a rotating spindle and a cutting tool, a coordinate measuring machine is equipped with electronic gaging probes which measure features on a workpiece. The results of the gaging are compared to acceptable limits programmed into the machine.

Robots are frequently used in an FMS to load and unload parts from the machines. Since robots are programmed pieces of equipment that lack the ability to make judgments, special workholding fixtures are employed on the transfer vehicles to orient the workpiece so that the robot can handle it correctly. Specially designed machine fixtures and clamping mechanisms are employed to ensure correct placement and clamping of the part on the machine. All part handling must be accomplished in a specific orderly fashion, with coordination of the part transfer vehicle, the robot, and the CNC machines. Future robots will probably employ some type of artificial intelligence which will enable them to make limited judgments as to workpiece orientation and take the necessary corrective actions.

The third critical component of an FMS is the tool and workpiece transfer vehicles. These vehicles shuttle workpieces from machine to machine. They also shuttle tool magazines to and from the machinery to maintain an adequate supply of sharp cutting tools at each CNC machine. Transfer vehicles employed in current flexible manufacturing systems are of four major types: automatic guided vehicles (AGV), wire guided vehicles, air cushion vehicles, and hardware guided vehicles.

Automatic guided vehicles rely on onboard sensors and/or a program to determine the path they take. There is no hardware connecting them to the system. An advantage of AGVs is that they can be reprogrammed to take different routes, eliminating the need to run tracks or wires for each route change. The corresponding disadvantage of AGVs is that they are the most difficult of the part delivery vehicles to make function, because of the lack of hardware connection.

A wire *guided vehicle* uses a wire buried in the floor to define its path. A sensor on the vehicle detects the location of the wire. A major advantage of wire guided vehicles is the ability to use the wire as opposed to an AGV without the need to have a hardware system such as an overhead wire or track on the floor. The disadvantage of wire guided vehicles is the necessity of installing new wire in the floor if a route change is required.

An air *cushion vehicle* is guided by some external hardware device, such as an overhead wire, but glides on a cushion of air rather than a track system. When using air cushioned vehicles, particular attention to chip removal and control must be built into the FMS. Chips in the path of an air cushion vehicle will stop its progress. These vehicles are generally used for straight paths.

Hardware guided vehicles are the most reliable but least flexible of the transfer vehicles. A track on the floor or an overhead guide rail controls the vehicle path. The advantages of these vehicles are their reliability and the ease of coordinating them with the rest of the system. The major disadvantage is, of course, the need to run new rail or track whenever a vehicle route change or new route is deemed necessary. A large FMS may employ several different types of vehicles, depending on the requirements of different parts of the manufacturing line.

FIGURE 16-2
(Photo couresty of Cincinnati Milacron)

Machining Cells

Large flexible machining systems are often a collection of smaller coordinated units called machining cells. A *machining cell* is a system consisting of one or more CNC machines and a parts handling device, such as a robot. The cell performs a machining operation or a specific sequence of operations. Flexible machining systems are not in widespread use at present although their numbers are increasing. Stand-alone machining cells, however, are widely employed by manufacturers, frequently with a view to incorporating them into an FMS at a later date. Figure 16-2 shows a machining cell consisting of a turning center and a grinder. A robot on an overhead gantry services the two machines.

EMPLOYMENT OPPORTUNITIES IN NC

A number of skilled positions have been created by numerical control. The most common jobs are NC electronics technician, machine operator/ setup operator, and part programmer.

Electronics Technician

Numerical control and computer numerical control equipment are electrical systems interfaced to a machine tool. The electronics necessary for a CNC machine to function are complex. The NC electronics technician is a skilled technician who specializes in the maintenance of numerical control equipment. The NC technician must be well trained in digital electronics and possess a knowledge of the cycles and functions of NC machinery. The technician must be able to troubleshoot and correct problems that occur in the electronic circuitry of various NC machines.

NC technicians generally acquire their skills through a two-year junior college program in digital electronics. Additional education in numerical control is often provided by the employer in the form of NC manufacturers' technical school classes and seminars.

Machine Operator/Setup Operator

The machine operator/setup operator is responsible for preparing an NC machine to run a program and for setting up the fixtures, tools, and workpieces. The operator must possess a knowledge of general machine shop practices and techniques, as well as the cycles and functions of an NC machine. The operator is responsible for overriding programmed speeds and feeds if required during machining. The operator also assigns the tool length offsets to the appropriate tool registers and may be called upon to single-step a program through its first cycle. The operator must also be trained in the use of precision measuring instruments as he or she is often responsible for measuring the parts as they are finished.

Machine operators/setup operators acquire their training either by years of running other types of manufacturing equipment and then transferring to an NC operator's position, or through a two-year junior college program. Factory seminars and other coursework may be provided by the employer as required.

Part Programmer

The part programmer is a highly skilled individual responsible for writing the programs that run on numerically controlled equipment. He or she must be trained in general machine shop practice, mathematics, and the use of computers. Based on the part drawing, the programmer selects a machine to machine the part and devises a machining strategy, listing the tools to be used and the coordinates necessary to accomplish the operations. This information is then assembled into a part program written for the particular machine selected.

An NC programmer may acquire training through a two-year junior college, a four-year engineering technology degree program, or by transferring from positions as journeyman machinists or tool and die makers. NC programmers take additional coursework and factory seminars as required by the employer. The educational requirements for a programmer vary with the employer.

SUMMARY

The important concepts presented in this chapter are:

- The use of CNC will increase in prototype and small job shops due to the arrival of lower cost controllers containing many advanced programming features.
- A flexible machining system is an unmanned manufacturing/assembly line that can take a part from raw stock and perform all the necessary operations to produce a finished part or assembly.
- A machining cell is a system of one or more CNC machines and part handling robots that performs a specific sequence of operations.
- An NC electronics technician is responsible for maintaining the electronics of an NC or CNC system.
- An NC operator/setup operator is responsible for preparing a machine prior to running a program and monitoring the machine during the program execution.
- An NC part programmer is responsible for creating the part program.

VOCABULARY INTRODUCED IN THIS CHAPTER

Air cushion vechicles
Flexible machining system (FMS)
Hardware guided vehicles
Machining cell
Robot
Wire guided vehicles

REVIEW QUESTIONS

1. For what reason will the use of CNC increase in one-of-a-kind and prototype shops?
2. What is a flexible machining system?
3. What are the four major components of an FMS?
4. What are the four types of part/tool transfer vehicles?
5. What is a machining cell?
6. What are the responsibilities of the CNC electronics technician?
7. What are the responsibilities of the CNC machine operator/setup operator?
8. What are the responsibilities of the CNC part programmer?

APPENDIX 1

EIA Codes

PREPARATORY FUNCTIONS

G00—Denotes rapid traverse for point-to-point positioning.

G01—Linear interpolation.

G02—Circular interpolation clockwise.

G03—Circular interpolation counterclockwise.

G04—Dwell.

G05–07—Unassigned.

G08—Acceleration at a smooth rate.

G09—Deceleration at a smooth rate.

G10–16—Unassigned.

G13–16—Axis selection codes.

G17—XY plane selection.

G18—ZX plane selection.

G19—YZ plane selection.

G20–32—Unassigned.

G33—Thread cutting, constant lead.

G34—Thread cutting, increasing lead.

G35—Thread cutting, decreasing lead.

G36–39—Unassigned.

G40—Cutter diameter compensation cancel.

G41—Cutter diameter compensation left.

G42—Cutter diameter compensation right.

G43—Cutter compensation inside corner (used to adjust for differences in programmed and actual cutter size).

G44—Cutter compensation outside corner (used to adjust for differences in programmed and actual cutter size).

G45–49—Unassigned.

G50–59—Used with adaptive controls.

G60–69—Unassigned.

G70—Inch programming.

G71—Metric programming.

G72—Three-dimensional circular interpolation clockwise.

G73—Three-dimensional circular interpolation counterclockwise.

G74—Multiquadrant circular interpolation cancel.

G75—Multiquadrant circular interpolation.
G76–79—Unassigned.
G80—Cycle cancel.
G81—Drill cycle.
G82—Drill cycle with dwell.
G83—Intermittent or deep hole drilling cycle.
G84—Tapping cycle.
G85–89—Boring cycles.
G90—Absolute positioning.
G91—Incremental positioning.
G92—Register preload code.
G93—Inverse time feedrate.
G94—Inches (millimeters) per minute feedrate.
G95—Inches (millimeters) per revolution feedrate.
G96—Unassigned.
G97—Revolutions per minute spindle speed.
G98–99—Unassigned.

MISCELLANEOUS FUNCTIONS

M00—Program stop.
M01—Optional (planned) stop.
M02—End of program.
M03—Spindle on clockwise.
M04—Spindle on counterclockwise.
M05—Spindle off.
M06—Tool change.
M07—Coolant on (flood).
M08—Coolant on (mist).
M09—Coolant off.
M10—Automatic clamp.
M11—Automatic unclamp.
M12—Synchronize multiple axes.
M13—Spindle clockwise and coolant on.
M14—Spindle counterclockwise and coolant on.
M15—Rapid motion positive direction.
M16—Rapid motion negative direction.
M17–18—Unassigned.
M19—Spindle orient and stop.
M20–29—Unassigned.
M30—End of tape, will rewind tape automatically.

M31—Interlock bypass.
M32–39—Unassigned.
M40–46—Gear changes if used, otherwise unassigned.
M47—Continues program execution from the start of program.
M48—Cancel M47.
M49—Deactivate manual speed or feed override.
M50–57—Unassigned.
M58—Cancel M59.
M59—RPM hold.
M60–99—Unassigned.

OTHER ADDRESSES

A—Rotary motion about the X axis.
B—Rotary motion about the Y axis.
C—Rotary motion about the Z axis.
D—Angular dimension around a special axis. Also used for a third feed function.
E—Angular dimension around a special axis, or special feed function.
H—Unassigned.
I—X axis arc centerpoint.
J—Y axis arc centerpoint.
K—Z axis arc centerpoint.
L—Unassigned.
O—Used on some controllers in place of N address for sequence numbers.
P—Special rapid traverse code, or a third axis parallel to the X axis.
Q—Special rapid traverse code, or a third axis parallel to the Y axis.
R—Special rapid traverse code, or a third axis parallel to the Z axis. Also used for radius designation.
U—Secondary axis parallel to X.
V—Secondary axis parallel to Y.
W—Secondary axis parallel to Z.

APPENDIX 2

Word Address Codes Used in Text Examples

PREPARATORY FUNCTIONS (G CODES) USED IN MILLING

Following is a list of preparatory functions used in CNC milling examples in this text. Other codes commonly used on General Numeric controllers are also listed.

G00—Rapid traverse positioning.

G01—Linear interpolation (feedrate movement).

G02—Circular interpolation clockwise.

G03—Circular interpolation counterclockwise.

G04—Dwell.

G10—Tool length offset value.

G17—Specifies X/Y plane.

G18—Specifies X/Z plane.

G19—Specifies Y/Z plane.

G20—Inch data input (on some systems).

G21—Metric data input (on some systems).

G22—Safety zone programming.

G23—Cross through safety zone.

G27—Reference point return check.

G28—Return to reference point.

G29—Return from reference point.

G30—Return to second reference point.

G40—Cutter diameter compensation cancel.

G41—Cutter diameter compensation left.

G42—Cutter diameter compensation right.

G43—Tool length compensation positive direction.

G44—Tool length compensation negative direction.

G45—Tool offset increase.

G46—Tool offset decrease.

G47—Tool offset double increase.

G48—Tool offset double decrease.

G49—Tool length compensation cancel.
G50—Scaling off.
G51—Scaling on.
G73—Peck drilling cycle.
G74—Counter tapping cycle.
G76—Fine boring cycle.
G80—Canned cycle cancel.
G81—Drilling cycle.
G82—Counter boring cycle.
G83—Peck drilling cycle.
G84—Tapping cycle.
G85—Boring cycle (feed return to reference level).
G86—Boring cycle (rapid return to reference level).
G87—Back boring cycle.
G88—Boring cycle (manual return).
G89—Boring cycle (dwell before feed return).
G90—Specifies absolute positioning.
G91—Specifies incremental positioning.
G92—Program absolute zero point.
G98—Return to initial level.
G99—Return to reference (R) level.

MISCELLANEOUS (M) FUNCTIONS USED IN MILLING AND TURNING

Following is a list of miscellaneous functions used in the milling and turning examples in this text. Other M functions common to General Numeric and FANUC controllers are also listed.

M00—Program stop.
M01—Optional stop.
M02—End of program (rewind tape).
M03—Spindle start clockwise.
M04—Spindle start counterclockwise.
M05—Spindle stop.
M06—Tool change.
M08—Coolant on.
M09—Coolant off.
M13—Spindle on clockwise, coolant on (on some systems).
M14—Spindle on counterclockwise, coolant on.
M17—Spindle and coolant off (on some systems).

M19—Spindle orient and stop.
M21—Mirror image X axis.
M22—Mirror image Y axis.
M23—Mirror image off.
M30—End of program, memory reset.
M41—Low range.
M42—High range.
M48—Override cancel off.
M49—Override cancel on.
M98—Jump to subroutine.
M99—Return from subroutine.

PREPARATORY FUNCTIONS (G CODES) USED IN TURNING

Following is a list of preparatory functions used in CNC milling examples in this text. Other codes commonly used on FANUC controllers are also listed.

G00—Rapid traverse positioning.
G01—Linear interpolation (feedrate movement).
G02—Circular interpolation clockwise.
G03—Circular interpolation counterclockwise.
G04—Dwell.
G10—Tool length offset value setting.
G17—Specifies X/Y plane.
G18—Specifies X/Z plane.
G19—Specifies Y/Z plane.
G20—Inch data input (on some systems).
G21—Metric data input (on some systems).
G22—Stored stroke limit on.
G23—Stored stroke limit off.
G27—Reference point return check.
G28—Return to reference point.
G29—Return from reference point.
G30—Return to second reference point.
G40—Tool nose radius compensation cancel.
G41—Tool nose radius compensation left.
G42—Tool nose radius compensation right.
G50—Programming of work coordinate system.
G68—Mirror image for double turrets on.
G69—Mirror image for double turrets off.

G70—Inch programming (some systems) or finish cycle.
G71—Metric programming (some systems) or stock removal in turning code.
G72—Stock removal in facing code.
G73—Pattern repeat.
G74—Z axis peck drilling.
G75—Groove cutting cycle, X axis.
G76—Multipass thread cutting.
G90—Absolute positioning.
G91—Incremental positioning.
G94—Per minute feed (some systems).
G95—Per revolution feed (some systems).
G98—Per minute feed (some systems).
G99—Per revolution feed (some systems).

APPENDIX 3

Codes in Common Use with Tape Machinery

PREPARATORY FUNCTIONS (g CODES)

(Note: On tape machinery, lowercase letters are generally used.)

g01—Linear interpolation.

g02—Circular interpolation clockwise.

g03—Circular interpolation counterclockwise.

g78—Mill cycle stop. A milling code used to position a spindle before lowering it. Upon receiving a g78 the spindle moves at rapid traverse to the programmed x/y coordinates, then rapids down to the feed engagement point, then feeds down to final depth at feedrate. The spindle is then clamped (either manually or automatically).

g79—Mill cycle. Usually (though not always) used following a g78. Upon receiving a g79 the spindle moves to the programmed x/y coordinates at feed rate, then rapids and subsequently feeds down to depth. If used following a g78, the spindle moves to the programmed coordinates at feedrate, since the spindle is already down.

g80—Cancel cycle.

g81—Drill cycle.

g84—Tapping cycle.

g85—Boring cycle.

MISCELLANEOUS (m) FUNCTIONS

m00—Program stop.

m02—End of program.

m03—Spindle on clockwise.

m04—Spindle on counterclockwise.

m05—Spindle off.

m06—Tool change.

m07—Flood coolant on.

m08—Mist coolant on.

m09—Coolant off.

m10—Clamp spindle.

m11—Spindle unclamp.

m13—Spindle on clockwise, flood coolant on.

m14—Spindle on counterclockwise, flood coolant on.

m17—Spindle on clockwise, mist coolant on.

m18—Spindle on counterclockwise, mist coolant on.

m26—Pseudo tool change. Used primarily for clamp changes. On some machines the spindle positions at rapid traverse to the tool change location but no tool change takes place. On other machines the spindle retracts to its "home" position.

m30—End of tape. Rewinds the tape to the start on machines where M02 does not do so.

m50–59—Z axis cam selection. Selects which of nine cams is to control z axis motion. m50 specifies no cam while m51–59 specify cams 1–9 respectively. Some machines use a 'w' function instead of an 'm' function to select cams.

APPENDIX 4

Safety Rules for Numerical Control

SAFETY RULES FOR OPERATING MACHINES

1. Use common sense in all situations.
2. Wear safety glasses at all times on the shop floor.
3. Wear safety shoes.
4. Keep long hair covered when operating or standing near a machine.
5. Do not wear jewelry (including rings), neckties, long sleeves, or loose clothing while operating machines.
6. Keep the floor free of obstructions.
7. Clean oil and grease spills immediately.
8. Do not play with compressed air or engage in general horseplay around machinery.
9. Do not use compressed air to clean machine slides. Chips blown under the machine ways will cause premature wear.
10. Do not perform grinding operations near NC machinery. Grinding grit will cause premature machine slide wear.
11. Platforms around machinery should be kept clean and have antislip surfaces.
12. Use caution when lifting heavy parts, tooling, or fixturing. Lift with the legs, not the back.
13. Keep tools and other parts off the machine.
14. Keep hands away from the spindle while it is revolving, and away from other moving parts of the machine.
15. Use a cloth or gloves when handling tools by their cutting edges.
16. Use caution when changing tools.
17. Use caution to avoid inadvertently bumping any NC controls.
18. Do not operate controls unless you have been instructed in their use.
19. Keep electrical panels in place. Electrical work should be performed only by qualified service personnel.
20. Make sure safety guards and devices are in place and working before operating the machine.

21. Do not remove chips from the machine or workpiece with hands or fingers. Use a brush. Do not remove chips with the spindle running.
22. Respect the programmer's knowledge of the machine.

SAFETY RULES FOR PROGRAMMERS

1. Never assume! When in doubt check the manual.
2. Do not attempt to program a machine without access to the programming manual for the machine tool and controller.
3. Cancel all modal commands in the first line of the program to ensure commands are not active when the program is cycled the first time.
4. Be sure all modal commands have been canceled at the end of the program so that no codes are active at the start of the next program.
5. Use a buffer zone between the part and the feed engagement point for all tool moves into the workpiece.
6. Respect the machine operator's knowledge of the machine.
7. When on the shop floor:
   a. Wear safety glasses at all times.
   b. Wear safety shoes.
   c. Remove neckties or tuck them inside your shirt.
   d. Keep hands away from moving machine parts.
   e. Keep long hair safely secured or covered.

APPENDIX 5

Useful Machining Formulas and Data

MACHINING FORMULAS

To determine spindle RPM:

$$RPM = \frac{(CS \times 4)}{D}$$

Where: CS is the material cutting speed in surface feet per minute, and D is the diameter of the part or cutter revolving in the spindle.

To determine feedrates:

    1. Milling feedrates:

$$FEED = RPM \times T \times N$$

Where: T is the chip load per tooth and N is the number of teeth on the cutter.

    2. Lathe feedrates (commonly .002–.025 inch per revolution):

$$FEED \ (in./rev) = I/RPM$$

Where: I is the feedrate in inches per minute.

$$FEED \ (in./min) = RPM \times r$$

Where: r is the feedrate in inches per revolution.

To determine lead of a thread:

$$LEAD = P \times I$$

Where: P is the pitch of the thread and I is the number of leads on the thread.

To determine pitch of a thread:

$$PITCH = 1/N$$

Where: N is the number of threads per inch.

To determine tap drill diameter of a thread (Unified threads):

$$MD - \left[\frac{1.08254 \times \%}{N} \right]$$

Where: MD is the major diameter of the thread, % is the percentage of thread engagement desired, and N is the number of threads per inch.

To determine length of a drill point:

$$\text{DRILL POINT} = .3 \times \text{DRILL DIAMETER}$$

To determine depth of countersink to achieve a given diameter:

$$\text{DEPTH} = A - (B \times C)$$

Where: A is the diameter of countersink desired, B is the diameter of the hole, and C is a constant as follows:

.35 for a 110-degree countersink
.50 for a 90-degree countersink
.57 for a 82-degree countersink
.35 for a 60-degree countersink

To determine circumference of a circle:

$$\text{CIRCUMFERENCE} = \text{DIAMETER} \times \text{PI}$$

To determine diameter of a circle:

$$\text{DIAMETER} = \text{CIRCUMFERENCE} \times .31831$$

To determine area of a circle:

$$\text{AREA} = \text{PI} \times \text{RADIUS}^2$$

$$\text{AREA} = \frac{1}{2}\,C \times \frac{1}{2}\,D$$

Where: C is the circumference and D is the diameter.

To determine surface area of a sphere:

$$\text{SURFACE} = \text{DIAMETER}^2 \times \text{PI}$$

To determine volume of a sphere:

$$\text{VOLUME} = \text{DIAMETER}^3 \times .5236$$

CUTTING SPEED DATA

The following rates are averages for high-speed steel cutters. For carbide cutters, double the cutting speed value.

Cutting speeds for lathes:

MATERIAL	CUTTING SPEED
Tool steel	50
Cast iron	60
Mild steel	100
Brass, soft bronze	200
Aluminum, magnesium	300

Cutting speeds for drills:

MATERIAL	CUTTING SPEED
Tool steel	50
Cast iron	60
Mild steel	100
Brass, soft bronze	200
Aluminum, magnesium	300

Cutting speeds for milling:

MATERIAL	CUTTING SPEED
Tool steel	40
Cast iron	50
Mild steel	80
Brass, soft bronze	160
Aluminum, magnesium	200

FEEDRATE DATA

Feeds for drilling:

DRILL SIZE	FEEDRATE
$< 1/8$	.001–.002
$1/8–1/4$	.002–.004
$1/4–1/2$	.004–.007
$1/2–1.000$	.007–.015
> 1.000	.025

Feeds per tooth for milling:

MATERIAL	FACE MILLS	SIDE MILLS	END MILLS
Low carbon steel	.010	.005	.005
Medium carbon steel	.009	.005	.004
High carbon steel	.006	.003	.002
Stainless steel	.006	.004	.002
Cast iron	.012	.006	.006
Brass and bronze	.013	.008	.006
Aluminum	.020	.012	.010

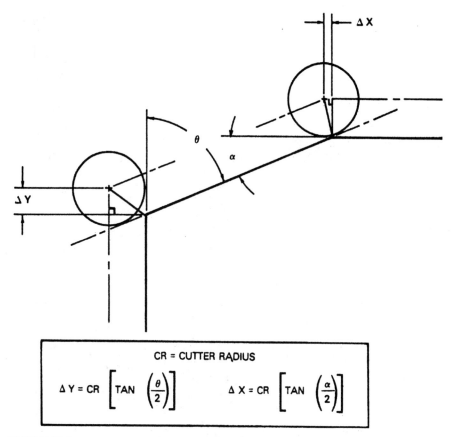

$$CR = CUTTER\ RADIUS$$

$$\Delta Y = CR\left[TAN\left(\frac{\theta}{2}\right)\right] \qquad \Delta X = CR\left[TAN\left(\frac{\alpha}{2}\right)\right]$$

FIGURE A5-1
Cutter offsets

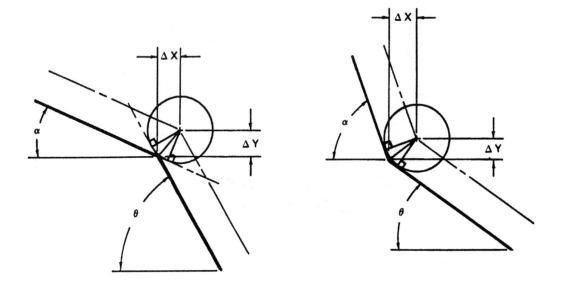

$$\Delta X = CR \times \frac{SIN\left(\dfrac{\alpha + \theta}{2}\right)}{COS\left(\dfrac{\alpha - \theta}{2}\right)} \qquad \Delta Y = CR \times \frac{COS\left(\dfrac{\alpha + \theta}{2}\right)}{COS\left(\dfrac{\alpha - \theta}{2}\right)}$$

FIGURE A5-2
Two lines intersecting, not parallel to a machine axis

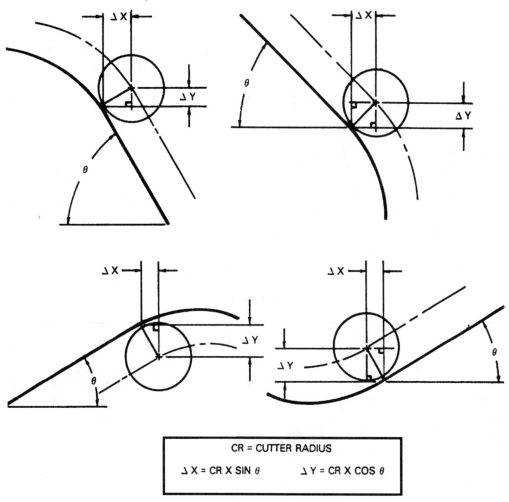

$$CR = CUTTER\ RADIUS$$
$$\Delta X = CR \times SIN\ \theta \qquad \Delta Y = CR \times COS\ \theta$$

FIGURE A5-3
Lines not parallel to a machine axis tangent to a circle

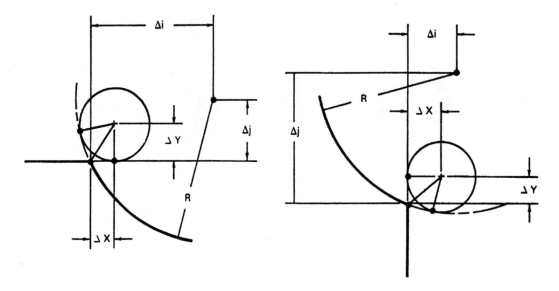

CR = CUTTER RADIUS

$$\lrcorner X = \Delta i - \sqrt{(R - CR)^2 - (\Delta j - CR)^2}$$

$$\lrcorner Y = CR$$

$$\lrcorner X = CR$$

$$\lrcorner Y = \Delta j - \sqrt{(R - CR)^2 - (\Delta i - CR)^2}$$

FIGURE A5-4
Intersection of a circle and a line parallel to a machine axis

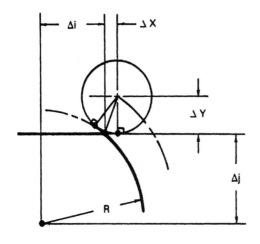

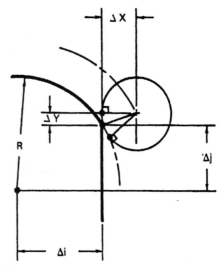

CR = CUTTER RADIUS

$$\Delta X = \sqrt{(R - CR)^2 - (\Delta j - CR)} - \Delta i$$
$$\Delta Y = CR$$

$$\Delta X = CR$$
$$\Delta Y = \sqrt{(R - CR)^2 - (\Delta i + CR)^2} - \Delta i$$

FIGURE A5-5
Intersection of a circle and a line parallel to a machine axis

APPENDIX 6

Lathe Canned Cycle Example

The program contained in this section is the courtesy of Hardinge Brothers Inc. Refer to Figures 1 and 2.

TOOLING USED

TURRET STATION	TOOL OFFSET	TOOLING USED
1	1	Number 4 centerdrill
3	3	55 deg. × .030r O.D. turning tool
4	4	Number 7 drill
5	5	1/4—20 tap
6	6	35 degrees × .015r O.D. turning tool
8	8	O.D. threading tool

SEQUENCE OF OPERATIONS

1. Centerdrill—1500 RPM, feedrate .009 IPR, .250 deep.
2. Rough face and turn—55 degree tool, G71 cycle, 400 constant surface speed with a maximum RPM of 4200, depth of cut .125, stock for finish pass .006, feedrate .008 IPR.
3. Tap drill—1900 RPM, feedrate .007 IPR, .875 deep.
4. Tap—250 RPM, G32 cycle, feedrate.049 IPR.
5. Finish face and turn—35 degree tool, G70 cycle, 450 constant surface speed with a maximum RPM of 5000, feedrate .004 IPM.
6. O.D. thread—G76 cycle, 1000 RPM, 8 passes used, 59 degree infeed angle.

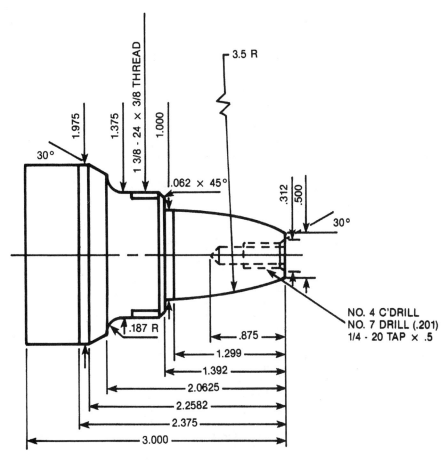

FIGURE A6-1
(Courtesy of Hardinge Brothers Inc.)

PROGRAM NOTES

The programmer of this part chose not to use sequence numbers on all lines. The uses of sequence numbers is up to the programmer, they are not required by the controller except where necessary for canned cycle or sub-routine use.

```
%
O401 (SAMPLE PART PROGRAM)
      G20
      G65 P9150 D1.5
N1    (T0101 #4 CENTER DRILL)
      G97 S1000 M13
      M98P1
      G4 T0101
      XO Z.100 S1500
      G99 Z-.25 F.009
      M98P2


N3    (T0303 ROUGH O.D. R.030 T3)
      G97  S1000 M13
      M98P1
      G4 T0303
      X2.16 Z.2
      G50 S4200
      G96 S400
      G42 X2.155 Z.1 F50.
      G99
      G71 P100 Q300 U.012 W.006 D12500 F.008
N100 G00 X.25
      G1 G99 ZO F.004
      X.5
      G3 X1. Z-1.299 R3.5
      G1 Z-1.302
      X1.375 K-.062
      Z-2.0625 R.1875
      X1.75
      X1.975 Z-2.2582
      Z-2.375
N300 X2.155
      M98P2


N4    (T0404 #7 DRILL)
      G97 S1000 M13
      M98P1
      G4 T0404
      XO Z.1 S1900
      G1 G99 Z-.875 F.007
      M98P2


N5    (T0505 1/4-20 TAP)
      G97 S250 M13
      M98P1
      G4 T0505
      XO Z.25
      G32 Z-.406 F.049   (TAPPED .094 SHORT FOR DRIVER PULLOUT)
      G4 U.5
      G32 Z.25 F.05 M14
      M98P2


N6    (T0808 1 3/8-24 O.D.  THREAD)
      G97 S1000 M13
      M98P1
      G4 T0808
      X1.4261 Z.1 S2880
      Z-1.4261 F50.
      G76 X1.3238 Z-1.767 K.0256 E.04166 D00904 A59
      M98P2
      TO100
      M30
      %
```

FIGURE A6-2
(Courtesy of Hardinge Brothers Inc.)

The uses of leading zeros is also optional at the programmer's discretion. In this example, G01 is written as G1, G02 as G2, etc.

0401
The "O" number. Each program resident in the controller must have its own unique "O" number.

G65 P9150 D1.5
Unique to Hardinge using FANUC controls. This code sets the dwell time parameter.

M98P1
Jump to subroutine call. This subprogram contains the startup code common to all tools. The subprogram is not part of the program printout.

M98P2
Jump to subroutine call. This subprogram contains the tool cancel code common to all tools. The subprogram is not part of the program printout.

CANNED CYCLE NOTES

G71 P100 Q300 U.012 W.006 D12500 F008
G71 is the rough face and rough turn cycle.

P100 is the sequence number (N100) that denotes the beginning of the finish pass lines.

Q300 is the sequence number (N300) that denotes the end of the finish pass lines.

U.012 tells the MCU to leave.012 stock on the X-axis for finishing. The lathe is diameter programmed, therefore, this will leave.006 stock per side to be finished.

W.006 tells the MCU to leave.006 stock on the Z-axis for finishing.

D12500 tells the MCU that the depth of each roughing pass is the.125 diameter stock removal.

F.008 is the feedrate.

The G71 cycle will calculate backwards from the finished pass lines to generate the appropriate number of passes. The tool path shape is identical to the finish pass.

G32 Z–.406 E049
G32—This type of cycle was covered in Chapter 14. It is the single pass threading cycle. When used for tapping, no X infeed is necessary. It is the same as using a G01 except with G32 the speed and feed operator overrides will not function.

F.049 is the feedrate. The tap is fed in at .001 IPR slower than the actual lead. This allows the tapping holder to pull out slightly. When the spindle is reversed at the bottom of the tapped hole, there is some travel in the holder to prevent tap breakage.

G70 P100 Q300

G70 is the finish face and turn cycle.
P100 is the sequence number (N100) of the finish pass lines.
Q300 is the sequence number (N300) of the finish pass lines.

G70 will simply jump backwards in the program to the finish pass lines used with the G71 cycle and execute them. Note the G42 prior to the G70 line. Tool nose radius comp (TNR) is being used to allow for tool wear.

G76 X1.3238 Z–1.767 K.0256 E.04166 D0904 A59

This cycle was covered in Chapter 14. The A59 sets the infeed angle to 59 degrees. Other programmers prefer to use A60.

APPENDIX 7

Sample Programs

SAMPLE PART PROGRAM

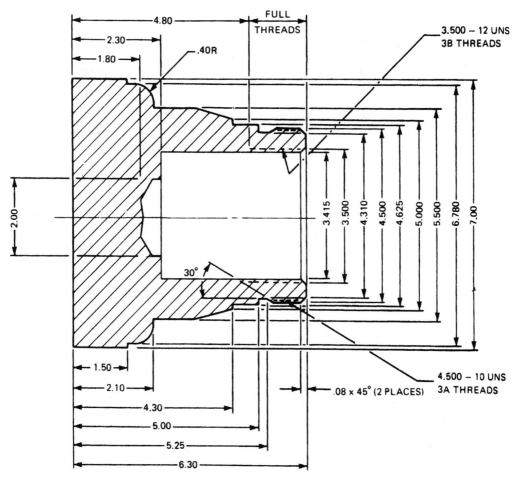

FIGURE A7-1
Sample Part Engineering Drawing *(Courtesy Cincinnati Milacron)*

OD TOOLING

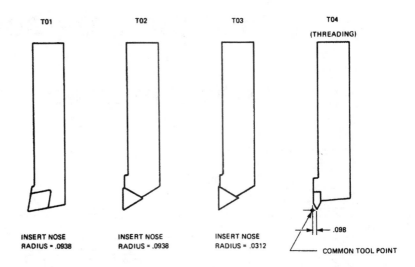

INSERT NOSE
RADIUS = .0938

INSERT NOSE
RADIUS = .0938

INSERT NOSE
RADIUS = .0312

.098

COMMON TOOL POINT

ID TOOLING

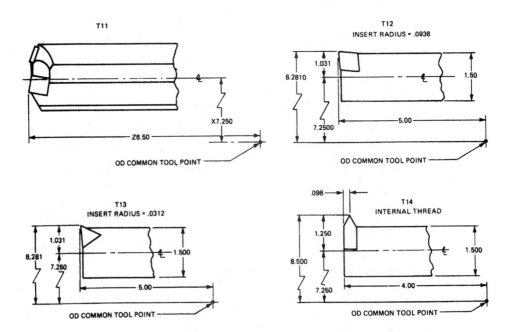

FIGURE A7-2
Tooling Layout

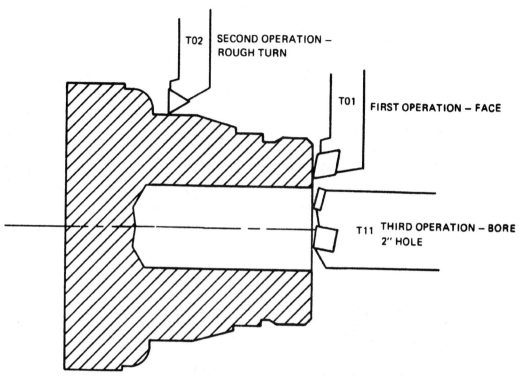

FIRST, SECOND AND THIRD OPERATIONS

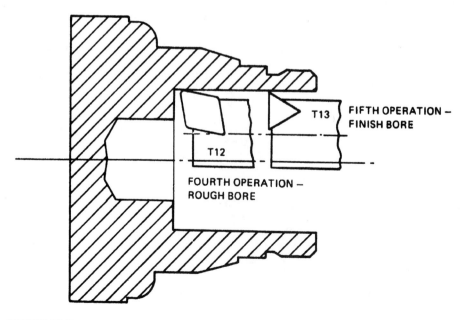

FIGURE A7-3

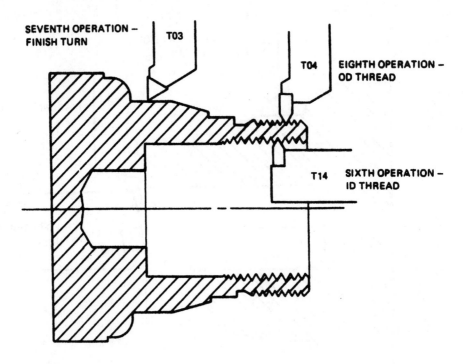

SIXTH, SEVENTH, AND EIGHTH OPERATIONS

FIGURE A7-4

SAMPLE PART PROGRAM NO. 1

First Operation   O10   G90

N20   G97 S100 M42

N30   G70 M03

N40   G00 X50000 Z85000 T0100 M06

N50   G95

N60   G92 S2500

N70   G96 R50000 S600

N80   G00 X37000 Z63500 M08

N90   G01 X−940 F150

N100   Z65500 F600

N110   X37000

N120   Z63000

N130   X−940 F150

N140   G00 Z65000

N150   X45000

Second Operation   O160   G90

N170   G97 S351 M41

N180   G70 M13

N190   G00 X45000 Z65000 T0200 M06

N200   G95

N210   G92 S2500

N220   G96 R45000 S600

N230   G00 X32600

N240   G01 Z19690 F150

N250   G03 X34100 Z16062 I28962 K16062

N260   G01 X37000

N270   G00 Z65000

N280   X30100

N290   G01 Z21200

N300   X32100

N310   G00 Z65000

N320   X27600

N330   G01 Z21100

N340   X28962

N350   G03 X34000 Z16062 I28962 K16062

N360   G01 Z15100

N370   X36000

N380   G00 X29600 Z65000

N390   X25100

N400   G01 Z42199

N410   X27600 Z32986

N420   G00 Z65000

N430   X22600

N440   G01 Z55989

N450   X21650 Z51659

N460   Z50100

N470   X23225

N480   Z43100

N490   X26000

N500   G00 Z130000

Third Operation   O510   G90

N520   G97 S600 M41

N530   G70 M14

N540   G00 X26000 Z130000 T1100 M06

FIGURE A7-5 *(Continues to page 400)*

```
                           N550   G95
                           N560   G00  X-72500
                           N570   G01  Z83000  F150
                           N580   G00  Z148000
     Fourth Operation      O590   G90  M05
                           N600   G97  S1000  M42
                           N610   G70  M13
                           N620   G00  X-72500  Z148000  T1200  M06
                           N630   G95
                           N640   G92  S2500
                           N650   G96  R10310  S600
                           N660   G00  X-71560  Z115000
                           N670   G01  Z73100  F150
                           N680   X-73750
                           N690   G00  Z115000
                           N700   X-70310
                           N710   G01  Z73100
                           N720   X-72500
                           N730   G00  Z115000
                           N740   X-69060
                           N750   G01  Z73100
                           N760   X-71250
                           N770   G00  Z115000
                           N780   X-67810
                           N790   G01  Z73100
                           N800   X-70000
                           N810   G00  Z115000
                           N820   X-66560
                           N830   G01  Z73100
                           N840   X-68750
                           N850   G00  Z115000
                           N860   X-65835
                           N870   G01  Z73100
                           N880   X-68023
                           N890   G00  Z148000
     Fifth Operation       O900   G90
                           N910   G97  S846  M42
                           N920   G70  M13
                           N930   G00  X-68023  Z148000  T1300  M06
                           N940   G95
                           N950   G92  S2500
                           N960   G96  R14787  S800
                           N970   G00  X-64752  Z115000
                           N980   G01  X-65735  Z112017  F100
                           N990   Z73000
                           N1000  X-73210
                           N1010  G00  Z148000
     Sixth Operation       O1020  G90  M05
                           N1030  G97  S400  M41
                           N1040  G70  M14
                           N1050  G00  X-73210  Z148000  T1400  M06
                           N1060  X-67805  Z103000
                           N1070  G91
                           N1080  G33  X-1500  Z-18668  K8333
                           N1090  G00  Z18613
                           N1100  X1600
```

```
                    N1110 G33 X−1500 Z−18668 K8333
                    N1120 G00 Z18629
                    N1130 X1570
                    N1140 G33 X−1500 Z−18668 K8333
                    N1150 G00 Z18640
                    N1160 X1550
                    N1170 G33 X−1500 Z−18668 K8333
                    N1180 G00 Z18646
                    N1190 X1550
                    N1200 G33 X−1500 Z−18668 K8333
                    N1210 G00 Z18654
                    N1220 X1525
                    N1230 G33 X−1500 Z−18668 K8333
                    N1240 G00 Z18670
                    N1250 X1520
                    N1260 G33 X−1500 Z−18668 K8333
                    N1270 G00
                    N1280 G90
                    N1290 Z105000
                    N1300 X45000
Seventh Operation   O1310 G90
                    N1320 G97 S780 M41
                    N1330 G70 M13
                    N1340 G00 X45000 Z105000 T0300 M06
                    N1350 G95
                    N1360 G92 S2500
                    N1370 G96 R45000 S800
                    N1380 G00 X19517 Z65000
                    N1390 G01 X22500 Z62017 F100
                    N1400 Z54373
                    N1410 X21550 Z52728
                    N1420 Z50000
                    N1430 X23125
                    N1440 Z43000
                    N1450 X24927
                    N1460 X27500 Z33399
                    N1470 Z21000
                    N1480 X29588
                    N1490 G03 X33900 Z16688 I29588 K16688
                    N1500 G01 Z15000
                    N1510 X37000
                    N1520 G00 X45000 Z65000
Eighth Operation    O1530 G90
                    N1540 G97 S300 M41
                    N1550 G70 M14
                    N1560 G00 X45000 Z65000 T0400 M06
                    N1570 X22300 Z66029
                    N1580 G33 Z50300 K10000
                    N1590 G00 X24500
                    N1600 Z65946
                    N1610 X22150
                    N1620 G33 Z50300 K10000
                    N1630 G00 X24500
                    N1640 Z65879
                    N1650 X22030
                    N1660 G33 Z50300 K10000
```

```
N1670 G00  X24500
N1680 Z65835
N1690 X21950
N1700 G33  Z50300  K10000
N1710 G00  X24500
N1720 Z65807
N1730 X21900
N1740 G33  Z50300  K10000
N1750 G00  X24500
N1760 Z65791
N1770 X21870
N1780 G33  Z50300  K10000
N1790 G00  X24500
N1800 Z65780
N1810 X21850
N1820 G33  Z50300  K10000
N1830 G00  X24500
N1840 X50000  Z85000
N1850 M30
```

FIGURE A7-5

SAMPLE MILLING PROGRAM

MAT: 103/1020 STEEL
6"x4" x 1" OR 1/2" THICK

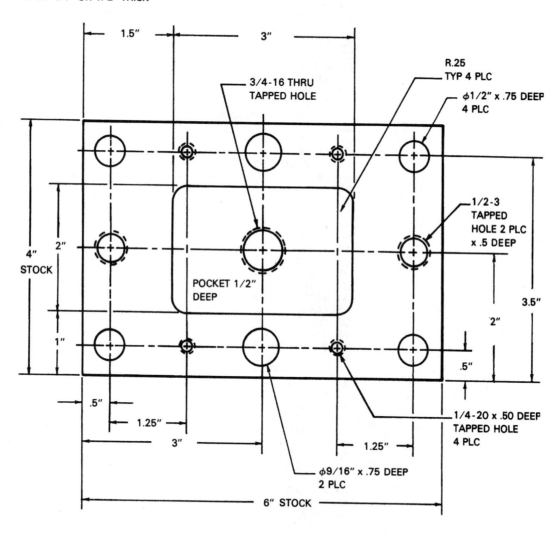

FIGURE A7-6
(Courtesy of Bayer Industries)

```
 1 O0001
 2 G20G40G49G80G90M03T01
 3 G00X-1.65Y2.75G0637
 4 G43Z0H01M08
 5 G01Z-.15F50.
 6 X6.25F22.93
 7 Y1.25
 8 X-1.65
 9 G00X0H00T02
10 X-1.65Y2.75S1490
11 G43Z0H02
12 G01Z-.16F50
13 X6.25F14.9
14 Y1.25
15 X-1.65
16 G00Z0H00T03
17 X3.Y2.S0407
18 G43Z0H03
19 G81G99Z-.725R0F6.11
20 G80Z0H00T04
21 X3.Y2.S0637
22 G43Z0H04
23 G01Z-.59F10.
24 G41X4.115F6.37D24
25 Y2.65
26 X2.385
27 Y1.35
28 X4.49
29 Y2.99
30 X1.51
31 Y1.01
32 X4.49
33 Z-.58F10.
34 G00G40X3.Y2.
35 Z0H00T05
36 X3.Y2.S1192
37 G43Z05H05
38 G01Z-.58F15.
39 Z-.6F11.92
40 G41X3.375D25
41 Y2.325
42 X2.625
43 Y1.675
44 X3.75
45 Y2.625
46 X2.25
47 Y1.375
48 X4.125
49 Y2.85
50 X1.875
51 Y1.15
52 X4.5
53 Y3
54 X1.5
```

FIGURE A7-7 *(Continues to page 404)*

```
55  Y1.
56  X4.5
57  Z-.59F15.
58  G00G40X3.Y2.
59  Z0H00T06
60  X3.Y2.80444
61  G43Z0H06
62  G81G98Z-1.25R-.5F6.66
63  G80Z0H00T07
64  X3.Y2.80162
65  G43Z0H07
66  G84G99Z-1.35R-.4F9.62
67  G80Z0H00T08
68  X.5Y.581222
69  G43Z0H08
70  G81G99Z-.25R0F6.11
71  X1.75
72  X3.
73  X4.25
74  X5.5
75  Y2.
76  Y3.5
77  X4.25
78  X3.
79  X1.75
80  X.5
81  Y2.
82  G80Z0H00T09
83  X.5Y.580611
84  G43Z0H009
85  G81G99Z-.75R0F7.33
86  Y3.5
87  X5.5
88  Y.5
89  G80Z0H00T10
90  X3.Y.580543
91  G43Z0H10
92  G81G99Z-.77R0F6.52
93  Y3.5
94  G80Z0H00T11
95  X1.75Y3.581520
96  G43Z0H11
97  G83G99Z-.66R0Q.25F9.12
98  X4.25
99  Y.5
100  X1.75
101  G80Z0H00T12
102  X1.75Y.580326
103  G43Z.2H12
104  G84G99Z-.4R.2F15.49
105  X4.25
106  Y3.5
107  X1.75
108  G80X0H00T13
```

```
109  X.5Y2.90724
110  843Z0H13
111  883899Z-.73R0Q.375F8.69
112  X5.5
113  880Z0H00T14
114  X5.5Y2.90255
115  843Z.2H14
116  884899Z-.45R.2F18.63
117  X.5
118  880Z0H00T15
119  X.5Y2.90795
120  843Z0H15
121  881899Z-.375R0F12.5
122  Y3.5
123  X1.75Z-.23
124  X3.Z-.4
125  X4.25Z-.23
126  X5.5Z-.375
127  Y2.
128  Y.5
129  X4.25Z-.23
130  X3.Z-.4
131  X1.75Z-.23
132  X.5Z-.375
133   880Z0H00M09
134  X-4.Y6.M05
135  M30
```

FIGURE A7-7

SAMPLE LATHE PROGRAM

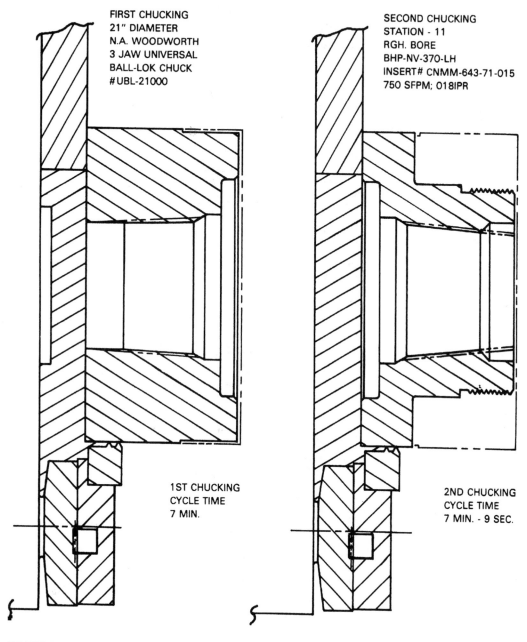

FIRST CHUCKING
21" DIAMETER
N.A. WOODWORTH
3 JAW UNIVERSAL
BALL-LOK CHUCK
#UBL-21000

SECOND CHUCKING
STATION - 11
RGH. BORE
BHP-NV-370-LH
INSERT# CNMM-643-71-015
750 SFPM; 018IPR

1ST CHUCKING
CYCLE TIME
7 MIN.

2ND CHUCKING
CYCLE TIME
7 MIN. - 9 SEC.

FIGURE A7-8
(Courtesy of Lodge & Shipley Co.)

```
N0010670M12
N00289T0202
N003692X15.Z22.782
N0048978065M03
N0050695
N006021X0.Z5.F.8
N007601Z-00.1F.007M08
N008821Z07.782F.8M09
N009X11.Z07.782
N0189T0101M12
N011692X06.7969Z20.2039S0712
N0128978021M04
N0130695
N014821X05.2Z05.182F.8
N0150896R5.1S06
N016001X1.7F.014M08
N0170Z5.212
N018621X04.95F.8
N0190601Z5.182F.022
N02603X05.057Z05.075K00.107
N0210601Z3.307
N022X05.15
N023821X06.7969Z20.2039F.8M09
N02489T0303
N0250M13
N026692X12.4531Z12.5789S15
N0278978071M04
N0280695
N0290621X1.9485Z5.295F.8
N0300896R2.9075
N031601Z-.1F.018M08
N03540X1.8485
N035021Z5.272F.8
N0360X2.2509
N037601Z4.047F.018
N0380X2.1058Z3.7366
N0382X2.0293Z1.2875
N0384Z-.1
N0386X1.9293
N0390621Z5.272F.8
N0400X2.4485
N0410601Z4.557F.018
N0420X2.1985
N0430621Z5.272F.8
N0440X2.6985
N0450601Z4.557F.018
N0460X2.4485
N0470621Z5.272F.8
N0480X2.9485
N0490601Z4.557F.018
N0500X2.4485
N0510621Z5.272F.8
N0520X3.1985
N0530601Z4.55F.018
```

FIGURE A7-9 *(Continues to page 409)*

```
N0540X2.9485
N0550G21Z5.272F.8
N0560X3.433
N0570G01Z4.635F.018
N575G03X03.355Z04.557I00.78
N0580G01X3.1985
N0590G21Z07.272F.8
N0592T0606
N0593G04X02.
N0595G92X03.1985Z07.272
N06G21X03.5406Z05.272
N0610G01Z5.172F.010
N0620X3.453Z5.0844
N0630Z4.625
N0640G03X03.375Z04.547I00.78
N0642G01X02.3209
N065X02.2609Z04.487
N0651Z04.047
N0652X02.0958Z03.7266
N0653X02.0458
N0660G97S0350M03
N069G21X01.4531Z07.272F.8M09
N0692G9T101
N0693G92X.967Z10.1941
N0694G21X2.0764Z5.5F.8
N0695G21X2.237M08
N0696G33X2.1209Z1.75I0039K.125
N0697G33X1.9597Z1.5888I.125K.125
N0698G21Z5.5F.8
N0699X2.2476
N0700G33X2.1309Z1.75I.0039K.125
N0701G33X1.9597Z1.5888I.125K.125
N0702G21Z8.F.8
N0703X11.967Z15.501
N0705G97S054M04
N071G9T0909M82
N0720
N073G92X06.781Z20.188S15
N074G97S054M04
N0750G95
N0760G21X3.48Z5.256F.8
N0770G96R3.581
N0780G01Z5.156F.008M08
N079X04.93
N08G03X05.021Z05.065K00.091
N0810G01Z3.281
N0820X5.2
N0825G97S038
N083G21X10.781Z35.188F.8M09
N0840T0000
N0845
N0850M00
N0855M12
N086G9T1101
```

```
N087G92X10.7969Z35.2039S0712
N088G97S021M004
N09G21X05.2Z05.057F.8
G0910G96R5.1S06
N092G01X1.8F.014M08
N0930Z5.147
N0940G21X4.567F.8
N0945M11
N0960G01Z1.807F.022
N097X04.95
N098G03X05.057Z01.7K00.107
N099G01X5.147F.8
N1G21Z05.147F.8
N1005M12
N1010X4.067
N1020G01Z1.807F.022
N1030X4.6
N1040G21Z5.147F.8
N0150X3.567
N1060G01Z1.807F.022
N1070X4.1
N1080G21Z5.147F.8
N0190X3.1644
N1100G01Z5.057F.022
N1110X3.317Z4.9094
N1120Z3.307
N1130X3.442
N1140Z1.885
N115G02X03.52Z01.1807I00.078
N1160G01X5.
N1170G97S023
N118G21X06.7969Z20.2039F.8M09
N119G9T2111
N1205M13
N121G92X12.4531Z16.0789S15
N122G97S07M04
N1230G95
N1240G21X2.3241Z5.147F.8
N125G96R2.3S075
N126G01Z04.422F.018M08
N1270X1.9465Z4.0444F.01
N1280G21Z5.147F.8
N1330X2.433
N1340G01Z4.432F.018
N1350X2.3
N136G21Z07.147F.8
N1362T0808
N1363G04X02.
N1365G92X02.3Z07.077
N137G21X02.56Z05.147
N1380G01Z5.04F.01
N139G02X02.453Z04.94K00.107
N1400G01Z4.422
N141X02.2741
```

```
N1420G21Z5.24F.8
N1425G97S076
N143X12.4531Z16.0089M09
N144G9T1909
N145G92Z06.781Z20.188S15
N146G97S076M04
N1470G95
N1480G21X2.48Z5.131F.8
N1490G96R2.581
N1500G01Z5.031F.01M08
N1510X3.2029
N1520X3.281Z4.9529
N1530Z3.281
N1540X3.345
N155G03X03.406Z03.22K00.061
N1560G01Z1.875
N157G02X03.5Z01.781I00.94
N158G01X04.93
N159G03X05.021Z01.69K00.91
N1600G01Z1.59
N1610X5.031
N16156G97S029
N162G21X06.781Z20.188F.8M09
N1621G9T0707
N1622G92X06.75Z20.728
N1623G97S035M04
N1624G95
N1625G21X03.4Z03.25F.8
N1626G96R03.4S06M04
N1627G01X03.125F.006M08
N1628G21X03.35F.8M8
N1629Z03.43
N163G01X03.17Z03.25F.006
N1631G21X03.4F.8
N1632G21X06.75Z10.728M09
N1634G9T0404
N164G92X06.75Z20.157
N1645M12
N1650G97S0290M03
N1660G95
N1670G21X3.35Z5.5F.8
N1600M08
N1690G83X-.122Z-.007H01
N1700G33Z3.375K.0625
N1710G84X-.007Z-.004H01
N1720G84X-.0055Z-.0032H01
N1730G84X-.0046Z-.0027H01
N1740G84X-.0041Z-.0024H01
N1750G84X-.0038Z-.0022H01
N1760G84X-.0034Z-.002H01
N1770G84X-.0Z-.0H01
N178G21X10.75Z35.157F.8M09
N1790T0000
N1795
N1800M30
```

FIGURE A7-9

GLOSSARY

The majority of this glossary is from Luggen, *Fundamentals of Numerical Control,* copyright 1984 by Delmar Publishers Inc. Reprinted with permission.

TERM AND DEFINITION	EXAMPLE

A AXIS   The axis of circular motion of a machine tool member or slide about the X axis. (Usually called alpha.)

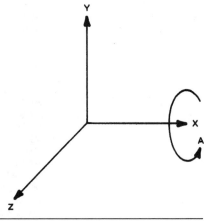

ABSOLUTE ACCURACY   Accuracy as measured from a reference which must be specified.

ABSOLUTE READOUT   A display of the true slide position as derived from the position commands within the control system.

ABSOLUTE SYSTEM   A numerical control system in which all positional dimensions, both input and feedback, are given with respect to a common datum point. The alternative is the incremental system.

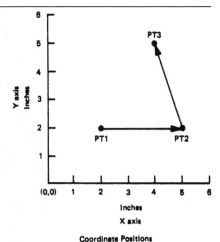

Coordinate Positions

Point	X value	Y value
PT1	2	2
PT2	5	2
PT3	4	5

In an absolute system, all points are relative to (0,0), and the absolute coordinates for each of the required points are programmed with respect to (0,0).

ACCURACY 1. Measured by the difference between the actual position of the machine slide and the position demanded. 2. Conformity of an indicated value to a true value, i.e., an actual or an accepted standard value. The accuracy of a control system is expressed as the deviation (the difference between the ultimately controlled variable and its ideal value), usually in the steady state or at sampled instants.

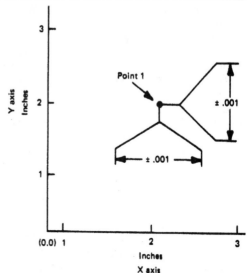

The position of point 1 in this example is X = 2 and Y = 2. If the machine accuracy is specified as ± .001, the X axis movement could be between X = 1.999 and X = 2.001. The Y axis movement could be between Y = 1.999 and Y = 2.001.

AD-APT An Air Force adaptation of APT program language with limited vocabulary. It can be used on some small to medium sizes of U.S. computers for NC programming.

C1 = circle/center, PT1, radius, 2.5

Similar to the APT language except it does not possess the advanced contouring capabilities of APT.

ADAPTIVE CONTROL A technique which automatically adjusts feeds and/or speeds to an optimum by sensing cutting conditions and acting upon them.

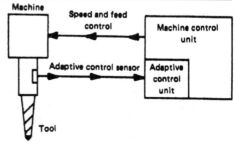

Sensors may measure variable factors, e.g. vibration, heat, torque, and deflection. Cutting speeds and feeds may be increased or decreased depending on conditions sensed.

ADDRESS 1. A symbol indicating the significance of the information immediately following. 2. A means of identifying information or a location in a control system. 3. A number which identifies one location in memory.

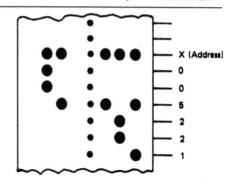

ALPHANUMERIC CODING A system in which the characters are letters A through Z and numerals 0 through 9.

APT and AD-APT statements use alphanumeric coding, e.g. GOFWD, CT12/PAST, 2, INTOF, L13

ANALOG 1. Applies to a system which uses electrical voltage magnitudes or ratios to represent physical axis positions. 2. Pertains to information which can have continuously variable values.

ANALYST A person skilled in the definition and development of techniques to solve problems.

APT (Automatic Programmed Tool) A universal computer-assisted program system for multiaxis contouring programming. APT III provides for five axes of machine tool motion.

Typical APT geometry definition statement:
C1 = CIRCLE/XLARGE, L12, XLARGE, L13, RADIUS, 3.5

Typical APT tool motion statement:
TLRGT, GORGT/AL3, PAST, AL12

ARC CLOCKWISE An arc generated by the co-ordinated motion of two axes, in which curvature of the tool path with respect to the workpiece is clockwise, when viewing the plane of motion from the positive direction of the perpendicular axis.

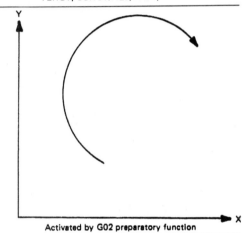

Activated by G02 preparatory function

ARC COUNTERCLOCKWISE An arc generated by the coordinated motion of two axes, in which curvature of the tool path with respect to the workpiece is counterclockwise, when viewing the plane of motion from the positive direction of the perpendicular axis.

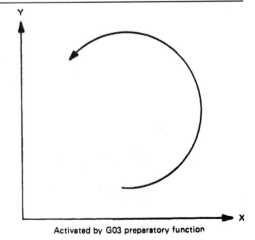

Activated by G03 preparatory function

ASCII (American Standard Code for Information Interchange) A data transmission code which has been established as an American standard by the American Standards Association. It is a code in which seven bits are used to represent each character. Formerly USASCII.

AUTO-MAP An abbreviation for AUTOmatic MAchining Programming. A computer aided programming language which is a subset of APT. It is used for simple contouring and straight line programming.

AUTOMATION 1. The implementation of processes by automatic means. 2. The investigation, design, development, and application of methods to render processes automatic, self-moving, or self-controlling.

AUTOSPOT (Automatic System for Positioning of Tools) A computer-assigned program for NC positioning and straight-cut systems, developed in the U.S. by the IBM Space Guidance Center. It is maintained and taught by IBM.

AUX CODE Auxiliary function command in Machinist Shop Language. Used to control specific functions within a CNC program.

AUXILIARY FUNCTION A programmable function of a machine other than the control of the coordinate movements or cutter.

- Transferring a tool to the select tool position.
- Turning coolant ON or OFF.
- Starting or stopping the spindle.
- Initiating pallet shuttle or movement.

AXIS A principal direction along which the relative movements of the tool or workpiece occur. There are usually three linear axes, mutually at right angles, designated as X, Y, and Z.

AXIS INHIBIT A feature of an NC unit which enables the operator to withhold command information from a machine tool slide.

AXIS INTERCHANGE The capability of inputting the information concerning one axis into the storage of another axis.

AXIS INVERSION The reversal of plus and minus values along an axis. This allows the machining of a left-handed part from right-handed programming or vice versa.

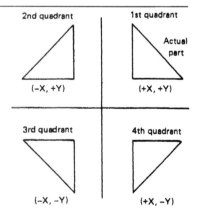

B (BETA) AXIS The axis of circular motion of a
machine tool member or slide about the Y axis.

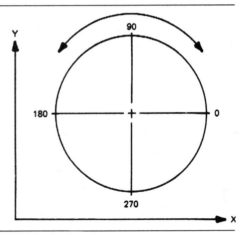

BACKLASH A relative movement between interacting mechanical parts as a result of looseness.

BCD (Binary-coded decimal) A system of number
representation in which each decimal digit is represented by a group of binary digits forming a
character.

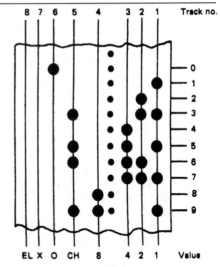

Numbers and letters are expressed by punched holes
across the tape for the code or value desired.

BINARY CODE   Based on binary numbers, which are expressed as either 1 or 0, true or false, on or off.

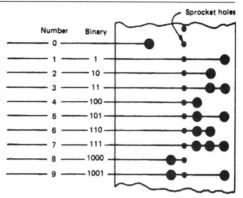

Most computers operate on some form of binary system where a number or letter can be expressed as ON (a hole) or OFF (no hole).

BIT   (Binary digit) 1. Binary digit having only two possible states. 2. A single character of a language using exactly two distinct kinds of characters. 3. A magnetized spot on any storage device.

BLOCK   A word, or group of words, considered as a unit. A block is separated from other units by an end of block character. On punched tape, a block of data provides sufficient information for an operation.

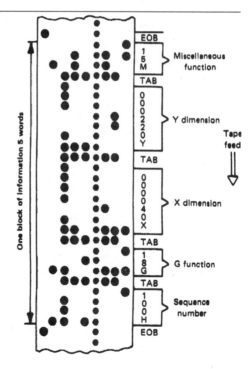

BLOCK DELETE Permits selected blocks of tape to be ignored by the control system, at the operator's discretion with permission of the programmer.

This feature allows certain blocks of information to be skipped by programming a slash (/) code in front of the block to be skipped. One lot of parts with holes 1, 2, and 3 are required. On another lot, only holes 1 and 3 are required. The same tape could be used for both lots by activating the block delete switch on the second lot and eliminating hole 2. The (/) code would be in front of the block of information for hole 2.

BUFFER STORAGE A place for storing information in a control system or computer for planned use. Information from the buffer storage section of a control system can be transferred almost instantly to active storage (that portion of the control system commanding the operation at the particular time). Buffer storage allows a control system to act immediately on stored information rather than wait for the information to be read into the machine from the tape reader.

BUG 1. A mistake or malfunction. 2. An integrated circuit (slang).

BYTE A sequence of adjacent binary digits usually operated on as a unit and shorter than a computer word.

Eight bits equal one byte. A computer word usually consists of either sixteen or thirty-two bits (two or four bytes).

CAD Computer-aided design.

CAM (Computer-aided manufacturing) The use of computers to assist in phases of manufacturing.

CAM-I (Computer Aided Manufacturing International) The outgrowth and replacement organization of the APT Long Range Program.

CANCEL A command which will discontinue any canned cycles or sequence commands.

CANNED CYCLE A preset sequence of events initiated by a single command. For example, code G84 will perform tap cycle by NC.

CARTESIAN COORDINATES A means whereby the position of a point can be defined with reference to a set of axes at right angles to each other.

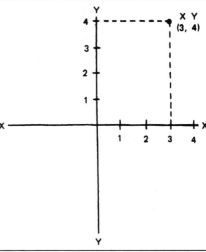

C AXIS Normally the axis of circular motion of a machine tool member or slide about the Z axis.

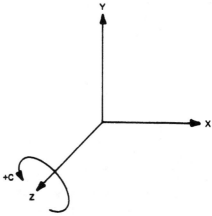

CHAD Pieces of material removed in card or tape operations.

CHANNELS Paths parallel to the edge of the tape along which information may be stored by the presence or absence of holes or magnetized areas. This term is also known as level or track. The EIA standard one-inch-wide tape has eight channels.

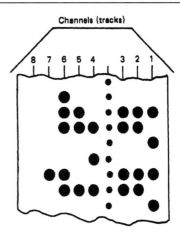

Channels (tracks)

CHARACTERS A general term for all symbols, such as alphabetic letters, numerals, and punctuation marks. It is also the coded representation of such symbols.

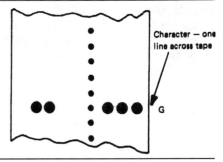

Character — one line across tape

G

CHIP A single piece of silicon cut from a slice by scribing and breaking. It can contain one or more circuits but is packaged as a unit.

CIRCULAR INTERPOLATION 1. Capability of generating up to 360 degrees of arc using only one block of information as defined by EIA. 2. A mode of contouring control which uses the information contained in a single block to produce an arc of a circle.

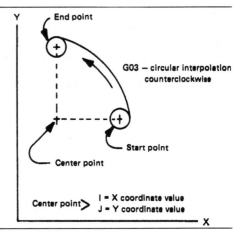

End point

G03 — circular interpolation counterclockwise

Start point

Center point

Center point> I = X coordinate value
J = Y coordinate value

CLOSED-LOOP SYSTEM A system in which the output, or some result of the output, is measured and fed back for comparison with the input. In an NC system, the output is the position of the table or head; the input is the tape information which ordinarily differs from the output. This difference is measured and results in a machine movement to reduce and eliminate the variance.

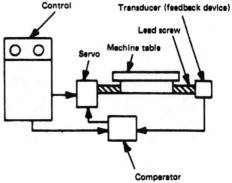

Control

Transducer (feedback device)

Lead screw

Machine table

Servo

Comparator

CNC Computer numerical control.

CODE A system describing the formation of characters on a tape for representing information, in a language that can be understood and handled by the control system.

COMMAND   A signal, or series of signals, initiating one step in the execution of a program.

COMMAND READOUT   A display of the slide position as commanded from the control system.

COMPUTER NUMERICAL CONTROL   A numerical control system utilizing an onboard computer as an MCU.

CONSTANT CUTTING SPEED   The condition achieved by varying the speed of rotation of the workpiece relative to the tool, inversely proportional to the distance of the tool from the center of rotation.

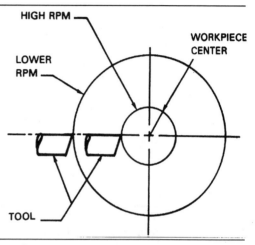

CONTINUOUS-PATH OPERATION   An operation in which rate and direction of relative movement of machine members is under continuous numerical control. There is no pause for data reading.

See *contouring control system.*

CONTOURING CONTROL SYSTEM   An NC system for controlling a machine (e.g., milling, drafting) in a path resulting from the coordinated, simultaneous motion of two or more axes.

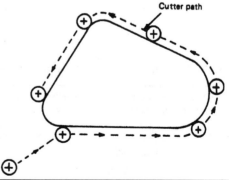

CPU   Central processing unit of a computer. The memory or logic of a computer that includes overall circuits, processing, and execution of instructions.

CRT   (Cathode Ray Tube) A device that represents data (alphanumeric or graphic) form by means of a controlled electron beam directed against a fluorescent coating in the tube.

CUTTER DIAMETER COMPENSATION A system in which the programmed path may be altered to allow for the difference between actual and programmed cutter diameters.

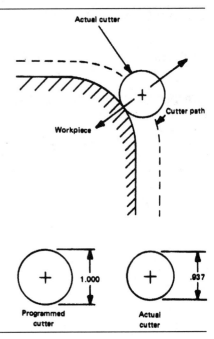

CUTTER OFFSET The distance from the part surface to the axial center of a cutter.

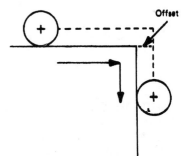

CUTTER PATH The path defined by the center of a cutter.

CYCLE 1. A sequence of operations that is repeated regularly. 2. The time it takes for one such sequence to occur.

DATA A representation of information in the form of words, symbols, numbers, letters, characters, digits, etc.

DATUM DIMENSIONING   A system of dimensioning based on a common starting point.

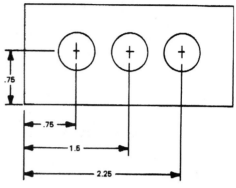

(Also known as absolute dimensioning)

DEBUG   1. To detect, locate, and remove mistakes from a program. 2. Troubleshoot.

DECIMAL CODE   A code in which each allowable position has one of ten possible states. (The conventional decimal number system is a decimal code.)

DELETE CHARACTER   A character used primarily to obliterate any erroneous or unwanted characters on punched tape. The delete character consists of perforations in all punching positions.

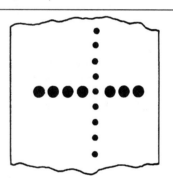

DELTA DIMENSIONING   A system used for defining part dimensions on a part drawing in which each dimension is referenced from the preceding one. Also known as incremental dimensioning in some shops.

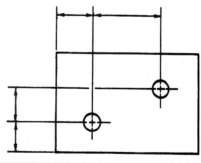

DIAGNOSTIC TEST   The running of a machine program or routine to discover a failure or potential failure of a machine element and to determine its location.

DIGIT   A character in any numbering system.

DIGITAL   1. Refers to discrete states of a signal (on or off). A combination of these makes up a specific value. 2. Relating to data in the form of digits.

DISPLAY   A visual representation of data.

DOCUMENTATION   Manuals and other printed materials (tables, magnetic tape, listing, diagrams) which provide information for use and maintenance of a manufactured product, both hardware and software.

DWELL   A timed or untimed delay in a program's execution. A timed dwell will resume the program after the programmed duration. An untimed dwell requires operator intervention to continue the program.

EDIT   To modify the form of data.

EIA STANDARD CODE   A standard code for positioning, straight-cut, and contouring control systems proposed by the U.S. EIA in their Standard RS-244. Eight-track paper (one-inch wide) has been accepted by the American Standards Association, as an American standard for numerical control.

END OF BLOCK CHARACTER   1. A character indicating the end of a block of tape information. Used to stop the tape reader after a block has been read. 2. The typewriter function of the carriage return when preparing machine control tapes.

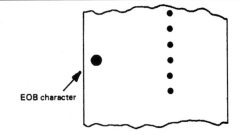

EOB character

END OF PROGRAM   A miscellaneous function (M02) indicating the completion of a workpiece. Stops spindle, coolant, and feed after completion of all commands in the block. Used to reset control and/or machine.

END OF TAPE   A miscellaneous function (M30) which stops spindle, coolant, and feed after completion of all commands in the block. Used to reset control and/or machine.

END POINT   The extremities of a span.

ERROR SIGNAL   Indication of a difference between the output and input signals in a servo system.

EXECUTIVE PROGRAM   A series of programming instructions enabling a dedicated minicomputer to produce a specific output control. For example, it is the executive program in a CNC unit that enables the control to think like a lathe or machining center.

FEED The programmed or manually established rate of movement of the cutting tool into the workpiece for the required machining operation.

FEEDBACK The transmission of a signal from a late to an earlier stage in a system. In a closed-loop NC system, a signal of the machine slide position is fed back and compared with the input signal, which specifies the demanded position. These two signals are compared and generate an error signal if a difference exists.

FEED FUNCTION The relative motion between the tool or instrument and the work due to motion of the programmed axis.

FEEDRATE (CODE WORD) A multiple-character code containing the letter F followed by digits. It determines the machine slide rate of feed.

FEEDRATE DIVIDER A feature of some machine control units that gives the capability of dividing the programmed feedrate by a selected amount as provided for in the machine control unit.

FEEDRATE MULTIPLIER A feature of some machine control units that gives the capability of multiplying the programmed feedrate by a selected amount as provided for in the machine control unit.

FEEDRATE OVERRIDE A variable manual control function directing the control system to reduce the programmed feedrate.

Feedrate override is a percentage function to reduce the programmed feed rate. If the programmed feed rate was 30 inches per minute and the operator wanted 15 inches per minute, the feedrate override dial would be set at 50 percent.

FIXED BLOCK FORMAT A format in which the number and sequence of words and characters appearing in successive blocks is constant.

FIXED CYCLE See canned cycle.

FIXED SEQUENTIAL FORMAT A means of identifying a word by its location in a block of information. Words must be presented in a specific order, and all possible words preceding the last desired word must be present in the block.

FLOATING ZERO A characteristic of a machine control unit permitting the zero reference point on an axis to be established readily at any point in the travel.

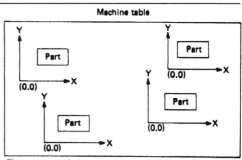

Machine table

The part or workpiece may be moved to *any* location on the machine table and zero may be established at that point.

FORMAT (TAPE)   The general order in which information appears on the input media, such as the location of holes on a punched tape or the magnetized areas on a magnetic tape.

FULL RANGE FLOATING ZERO   A characteristic of a numerical machine tool control permitting the zero point on an axis to be shifted readily over a specified range. The control retains information on the location of permanent zero.

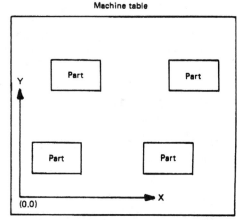

The part or workpiece may be shifted to any position on the machine table, but the actual position of permanent zero remains constant.

GAGE HEIGHT   A predetermined partial retraction point along the Z axis to which the cutter retreats from time to time to allow safe XY table travel. Also called the reference or rapid level.

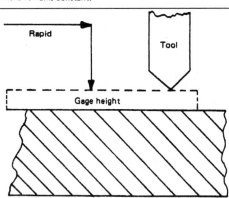

Gage height, usually .100 to .125, is a set distance established in the control or set by the operator. Gage height allows the tool, while advancing in rapid traverse, to stop at the established distance (gage height) and begin feed motion. Without gage height, the tool would rapid into the part causing tool damage or breakage and potential operator injury.

G CODE   A word addressed by the letter G and followed by a numerical code defining preparatory
functions or cycle types in a numerical control
system.

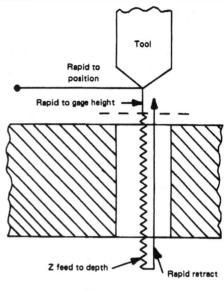

G81 — Drill Cycle

GENERAL PROCESSOR   1. A computer program for converting geometric input
data into cutter path data required by an NC machine. 2. A fixed software program designed for a specific logical manipulation of data.

HARD COPY   A readable form of data output on paper.

HARDWARE   The component parts used to build a computer or control system, e.g.
integrated circuits, diodes, transistors.

HARD-WIRED   Having logic circuits interconnected on a backplane to give a fixed
pattern of events.

HIGH-SPEED READER   A reading device which can be connected to a computer
or control so as to operate on line without seriously holding up the computer or
control.

INCREMENTAL SYSTEM   A control system in which each coordinate or positional dimension, both input and feedback, is taken from the past position rather than from a common datum point, as in the absolute system.

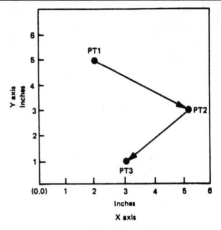

Coordinate positions

Point	X value	Y value
PT1	2	5
PT2	3	-2
PT3	-2	-2

In an incremental system, all points are expressed relative to the preceding point.

INDEX TABLE   A multiple-character code containing the letter B followed by digits. This code determines the position of the rotary index table in degrees.

See B (Beta) axis.

INHIBIT   To prevent an action or acceptance of data by applying an appropriate signal to the appropriate input.

INITIAL LEVEL   The position of the spindle at the beginning of a canned cycle operation.

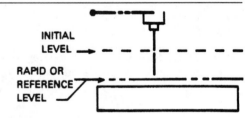

INPUT   Transfer of external information into the control system.

INPUT MEDIA   1. The form of input such as punched cards and tape or magnetic tape. 2. The device used to input information.

INTERCHANGEABLE VARIABLE BLOCK FORMAT   A programming arrangement consisting of a combination of the word address and tab sequential formats to provide greater compatibility in programming. Words are interchangeable within the block. Length of block varies since words may be omitted.

This is one of the most sophisticated tape formats in use today.

See *block*.

INTERCHANGE STATION   The position where a tool of an automatic tool changing machine awaits automatic transfer to either the spindle or the appropriate coded drum station.

INTERMEDIATE TRANSFER ARM   The mechanical device in automatic tool changing that grips and removes a programmed tool from the coded drum station and places it into the interchange station, where it awaits transfer to the machine spindle. This device then automatically grips and removes the used tool from the interchange station and returns it to the appropriate coded drum station.

INTERPOLATION   1. The insertion of intermediate information based on an assumed order or computation. 2. A function of a control whereby data points are generated between given coordinate positions.

INTERPOLATOR   A device which is part of a numerical control system and performs interpolation.

ISO   International Organization for Standardization.

JOG   A control function which momentarily operates a drive to the machine.

LEADING ZEROES   Redundant zeroes to the left of a number.

Leading zeroes

$$X + \overset{\frown}{00}62500$$

LEADING ZERO SUPPRESSION   See zero suppression.

LETTER ADDRESS   The method by which information is directed to different parts of the system. All information must be preceded by its proper letter address, e.g., X, Y, Z, M.

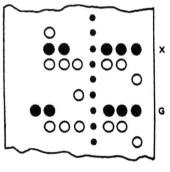

X and G address

An identifying letter inserted in front of each word.

LINEAR INTERPOLATION A function of a control whereby data points are generated between given coordinate positions to allow simultaneous movement of two or more axes of motion in a linear (straight) path.

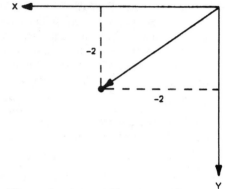

The control system moves X and Y axes proportionately to arrive at the destination point.

LOOP TAPE A short piece of tape, with joined ends, which contains a complete program or operation.

MACHINING CENTER Machine tools, usually numerically controlled, capable of automatically drilling, reaming, tapping, milling, and boring multiple faces of a part. Equipped with a system for automatically changing cutting tools.

MACRO A group of instructions which can be stored and recalled as a group to solve a recurring problem.

An APT macro could be as follows:

```
DRILL1 = MACRO/X, Y, Z, Z1, FR, RR
     GOTO/POINT, X, Y, Z, RR
     GODLTA/-Z1, FR
     GODLTA/+Z1, RR
TERMAC
```

X, Y, Z, Z1, FR, and RR would be variables which would have values assigned when the macro is called into action. The variables would be as follows:

- X = X position
- Y = Y position
- Z = Z position (above work surface)
- Z1 = Z feed distance
- FR = feed rate
- RR = rapid rate

The call statement could be:

```
CALL/DRILL1, X = 2, Y = 4, Z = .100, Z1 = 1.25,
     FR = 2, RR = .200
```

MAGIC-THREE CODING A feedrate code that uses three digits of data in the F word. The first digit defines the power of ten multiplier. It determines the positioning of the floating decimal point. The last two digits are the most significant digits of the desired feedrate.

To program a feed rate of 12 inches per minute in magic-three coding:

1) count the number of decimal places to the left of the decimal. .12 = 2
2) Add magic "3" to the number of counted decimal places. (3 + 2 = 5)
3) write the F word address, the added digit, and the first two digits of the actual feed rate to be programmed. (F512)
4) F512 would be the magic "3" coded feed rate.

This method of feed rate coding is now almost obsolete.

MAGNETIC TAPE A tape made of plastic and coated with magnetic material. It stores information by selective polarization of portions of the surface.

MANUAL DATA INPUT A mode or control that enables an operator to insert data into the control system. The data are identical to information that could be inserted by tape.

MANUAL PART PROGRAMMING The preparation of a manuscript in machine control language and format to define a sequence of commands for use on an NC machine.

Manual, or hand, programming is programming the actual codes, X and Y positions, functions, etc. as they are punched in the N/C tape.

H001 G81   X+37500 Y+52500   W01

MANUSCRIPT A written or printed copy, in symbolic form, containing the same data as that punched on cards or tape or retained in a memory unit.

MEMORY An organized collection of storage elements, e.g., disc, drum, ferrite cores, into which a unit of information consisting of a binary digit can be stored and from which it can later be retrieved.

A computer with a 64,000-word capacity is said to have a memory of 64 K.

MIRROR IMAGE See axis inversion.

MODAL Information that is retained by the system until new information is obtained and replaces it.

MODULE An interchangeable plug-in item containing components.

NC (Numerical control) The technique of controlling a machine or process by using command instructions in coded numerical form.

NULL 1. Pertaining to no deflection from a center or end position. 2. Pertaining to a balanced or zero output from a device.

NUMERICAL CONTROL SYSTEM A system in which programmed numerical values are directly inserted, stored on some form of input medium, and automatically read and decoded to cause a corresponding movement in a machine or process.

OFFLINE PROGRAMMING The development of an NC part program away from the machine console to be transferred to the MCU at a later time.

OFFSET A displacement in the axial direction of the tool which is the difference between the actual tool length and the programmed tool length.

OPEN-LOOP SYSTEM   A control system that has no means of comparing the output with the input for control purposes. No feedback.

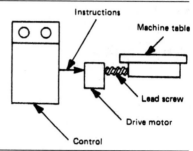

OPTIMIZE   To rearrange the instructions or data in storage so that a minimum number of transfers are required in the running of a program. To obtain maximum accuracy and minimum part production time by manipulation of the program.

OPTIONAL STOP A miscellaneous function (M01) command similar to Program Stop except the control ignores the command unless the operator has previously pushed a button to validate the command.

OVERSHOOT   A term applied when the motion exceeds the target value. The amount of overshoot depends on the feedrate, the acceleration of the slide unit, or the angular change in direction.

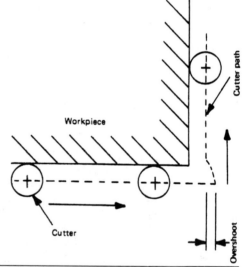

PARABOLA   A plane curve generated by a point moving so that its distance from a fixed second point is equal to its distance from a fixed line.

PARABOLIC INTERPOLATION   Control of cutter path by interpolation between three fixed points by assuming the intermediate points are on a parabola.

PARITY CHECK   1. A hole punched in one of the tape channels whenever the total number of holes is even, to obtain an odd number, or vice versa depending on whether the check is even or odd. 2. A check that tests whether the number of ones (or zeroes) in any array of binary digits is odd or even.

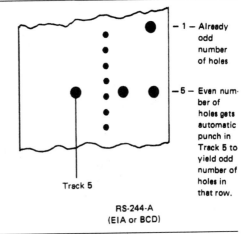

Track 5

- 1 – Already odd number of holes

- 5 – Even number of holes gets automatic punch in Track 5 to yield odd number of holes in that row.

RS-244-A
(EIA or BCD)

PART PROGRAM   A specific and complete set of data and instructions written in source languages for computer processing or in machine language for manual programming to manufacture a part on an NC machine.

PART PROGRAMMER   A person who prepares the planned sequence of events for the operation of a numerically controlled machine tool.

PERFORATED TAPE   A tape on which a pattern of holes or cuts is used to represent data.

PLOTTER   A device which will draw a plot or trace from coded NC data input.

POINT-TO-POINT CONTROL SYSTEM   A numerical control system in which controlled motion is required only to reach a given end point, with no path control during the transition from one end point to the next.

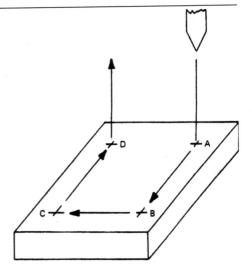

POSITIONING/CONTOURING   A type of numerical control system that has the capability of contouring, without buffer storage, in two axes and positioning in a third axis for such operations as drilling, tapping, and boring.

POSITIONING SYSTEM   See point-to-point control system.

POSITION READOUT   A display of absolute slide position as derived from a position feedback device (transducer) normally attached to the lead screw of the machine. See command readout.

POSTPROCESSOR   The part of the software which converts the cutter path coordinate data into a form which the machine control can interpret correctly. The cutter path coordinate data are obtained from the general processor and all other programming instructions and specifications for the particular machine and control.

PREPARATORY FUNCTION   An NC command on the input tape changing the mode of operation of the control. (Generally noted at the beginning of a block by the letter G plus two digits.)

Some preparatory functions are:

G84 — tap cycle
G01 — linear interpolation
G82 — dwell cycle
G02 — circular interpolation — clockwise
G03 — circular interpolation — counter clockwise

See *G code.*

PROGRAM   A sequence of steps to be executed by a control or a computer to perform a given function.

PROGRAMMED DWELL   The capability of commanding delays in program execution for a programmable length of time.

PROGRAMMER (PART PROGRAMMER)   A person who prepares the planned sequence of events for the operation of a numerically controlled machine tool. The programmers principal tool is the manuscript on which the instructions are recorded.

Manual part programming instructions:

```
H001  G81   X+123750   Y+62500   W01
N002        X+105000
N003                    Y+51250              M06
```

Computer part programming instructions:

```
TLRGT,  GORGT/HL3, TANTO, C1
        GOFWD/C1, TANTO, HL2
        GOFWD/HL2, PAST, VL2
```

PROGRAM STOP   A miscellaneous function (M00) command to stop the spindle, coolant, and feed after completion of the dimensional move commanded in the block. To continue with the remainder of the program, the operator must push a button.

QUADRANT Any of the four parts into which a plane is divided by rectangular coordinate axes in that plane.

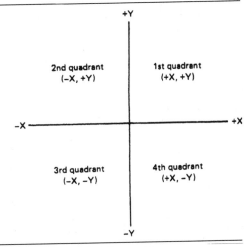

RANDOM Not necessarily in a logical order of arrangement according to usage, but having the ability to select from any location and in any order from the storage system.

RAPID Positioning the cutter and workpiece into close proximity with one another at a high rate of travel speed, usually 150 to 400 inches per minute (IPM) before the cut is started.

READER A pneumatic, photoelectric, or mechanical device used to sense bits of information on punched cards, punched tape, or magnetic tape.

REGISTER An internal array of hardware binary circuits for temporary storage of information.

REPEATABILITY Closeness of, or agreement in, repeated measurements of the same characteristics by the same method, using the same conditions.

RESET To return a register or storage location to zero or to a specified initial condition.

ROW (TAPE) A path perpendicular to the edge of the tape along which information may be stored by the presence or absence of holes or magnetized areas. A character would be represented by a combination of holes.

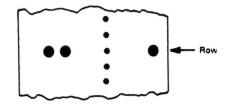

SEQUENCE NUMBER (CODE WORD) A series of numerals programmed on a tape or card and sometimes displayed as a readout; normally used as a data location reference or for card sequencing.

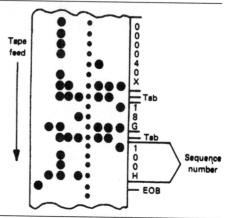

SEQUENCE READOUT A display of the number of the block of tape being read by the tape reader.

SEQUENTIAL Arranged in some predetermined logical order.

SIGNIFICANT DIGIT A digit that must be kept to preserve a specific accuracy or precision.

Significant digits

$$X + \underbrace{00}\underbrace{52500}$$

Insignificant digits

SLOW-DOWN SPAN A span of information having the necessary length to allow the machine to decelerate from the initial feedrate to the maximum allowable cornering feedrate that maintains the specified tolerance.

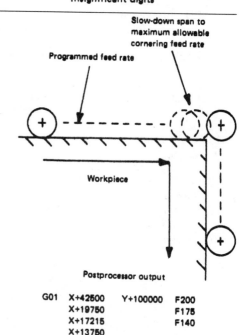

Postprocessor output

G01	X+42500	Y+100000	F200
	X+19750		F175
	X+17215		F140
	X+13750		
		Y+68750	F200

SOFTWARE   Instructional literature and computer programs used to aid in part programming, operating, and maintaining the machining center.

Examples of software programs are:

APT
FORTRAN
COBOL
RPG

SPAN   A certain distance or section of a program designated by two end points for linear interpolation; a beginning point, a center point, and an ending point for circular interpolation; and two end points and a diameter point for parabolic interpolation.

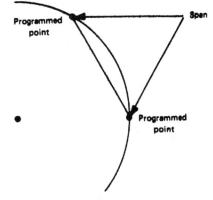

One linear interpolation span

SPINDLE SPEED (CODE WORD)   A multiple-character code containing the letter S followed by digits. This code determines the RPM of the cutting spindle of the machine.

STORAGE   A device into which information can be introduced, held, and then extracted at a later time.

STORAGE MEDIA   A device onto which information can be transferred and retained for later use. Storage media may also be used as input media, thereby serving a dual purpose.

TAB   A nonprinting spacing action on tape preparation equipment. A tab code is used to separate words or groups of characters in the tab sequential format. The spacing action sets typewritten information on a manuscript into tabular form.

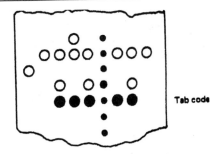

Tab code

TAB SEQUENTIAL FORMAT   Means of identifying a word by the number of tab characters preceding the word in a block. The first character of each word is a tab character. Words must be presented in a specific order, but all characters in a word, except the tab character, may be omitted when the command represented by that word is not desired.

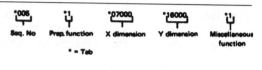

*005	*1	*07000	*16000	*1
Seq. No	Prep. function	X dimension	Y dimension	Miscellaneous function

* = Tab

The tab sequential format is, for the most part, obsolete.

TAPE   A magnetic or perforated paper medium for storing information.

TAPE LAGGER   The trailing end portion of a tape.
TAPE LEADER   The front or lead portion of a tape.

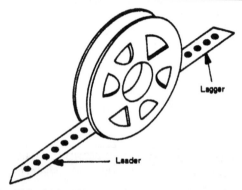

Lagger

Leader

Reel tapes should have a leader and lagger of approximately three feet with just sprocket holes for tape loading and threading purposes.

TOOL FUNCTION   A tape command identifying a tool and calling for its selection. The address is normally a T word.

T06 would be a tape command calling for the tool assigned to spindle or pocket 6 to be put in the spindle.

TOOL LENGTH COMPENSATION   A manual input, by means of selector switches, to eliminate the need for preset tooling; allows the programmer to program all tools as if they are of equal length.

TOOL OFFSET   1. A correction for tool position parallel to a controlled axis. 2. The ability to reset tool position manually to compensate for tool wear, finish cuts, and tool exchange.

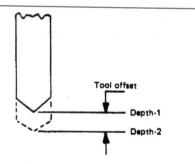

Tool offset

Depth-1

Depth-2

Tool offsets are used as final adjustments to increase or decrease depths due to cutting forces and tool deflection In this case, a tool offset could be used to increase the drill depth from depth-1 to depth-2.

TRAILING ZERO SUPPRESSION See zero suppression.

TURNKEY SYSTEM A term applied to an agreement whereby a supplier will install an NC or computer system so that he has total responsibility for building, installing, and testing the system.

USASCII United States of America Standard Code for Information Interchange. See ASCII.

VARIABLE BLOCK FORMAT (TAPE) A format which allows the quantity of words in successive blocks to vary. Same as word address. Variable block means the length of the blocks can vary depending on what information needs to be conveyed in a given block. See *block.*

VECTOR A quantity that has magnitude, direction, and sense; is represented by a directed line segment whose length represents the magnitude and whose orientation in space represents the direction.

VECTOR FEEDRATE The feedrate at which a cutter or tool moves with respect to the work surface. The individual slides may move slower or faster than the programmed rate, but the resultant movement is equal to the programmed rate.

WORD An ordered set of characters which is the normal unit in which information may be stored, transmitted, or operated upon.

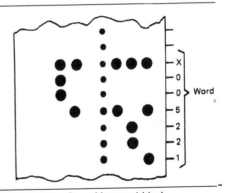

See *address* and *block.*

WORD ADDRESS FORMAT The specific arrangement of addressing each word in a block of information by one or more alphabetical characters which identify the meaning of the word.

WORD LENGTH The number of bits or characters in a word. See *word.*

X AXIS Axis of motion that is always horizontal and parallel to the workholding surface.

Y AXIS Axis of motion that is perpendicular to both the X and Z axes.

Z AXIS Axis of motion that is always parallel to the principal spindle of the machine.

ZERO OFFSET   A characteristic of a numerical machine tool control permitting the zero point on an axis to be shifted readily over a specified range. The control retains information on the location of the permanent zero.

See *full range floating zero* and *floating zero.*

ZERO SHIFT A characteristic of a numerical machine tool control permitting the zero point on an axis to be shifted readily over a specified range. (The control does not retain information on the location of the permanent zero.)

See *floating zero.* Consult chapter 4 for additional details.

ZERO SUPPRESSION   Leading zero suppression: the elimination of insignificant leading zeroes to the left of significant digits usually before printing. Trailing zero suppression: the elimination of insignificant trailing zeroes to the right of significant digits usually before printing.

Leading zero suppression

X + 0043500

Insignificant digits

Could be written as:

X + 43500

Trailing zero suppression

X + 0043500

Insignificant digits

Could be written as:

X + 00435

RELATED SME TITLES

INDEX